ErgoWise

ErgoWise

A Personal Guide to Making Your Workspace Comfortable and Safe

William A. Schaffer and Rab Cross, M.D., C.P.E.

amacom
American Management Association
New York • Atlanta • Boston • Chicago • Kansas City • San Francisco • Washington, D.C.
Brussels • Mexico City • Tokyo • Toronto

This book is available at a special discount when ordered in bulk quantities. For information, contact Special Sales Department, AMACOM, a division of American Management Association, 135 West 50th Street, New York, NY 10020.

Library of Congress Cataloging-in-Publication Data

Schaffer, William A.
ErgoWise : a personal guide to making your workspace comfortable and safe / William A. Schaffer and Rab Cross.
p. cm.
Includes index.
ISBN 0-8144-7907-3
1. Human engineering. 2. Office furniture. 3. Office layout. 4. Work environment. I. Cross, Rab. II. Title.
TA166.S36 1996
620.8—dc20 *96-4888*
CIP

Illustrations by Susan Detrich

Printing number

10 9 8 7 6 5 4 3 2 1

To the legions of women and men who put in long hours in offices, in every kind of organization, as well as to those who work at home, we respectfully dedicate this book.

Contents

Preface

ErgoWise is designed to provide information that will enable you to understand your workplace better, so that you can take charge of your working environment and thus improve your safety and comfort in it.

Chapter 1 introduces you to the office environment and some of the risks it may present in the form of repetitive strain injuries. It explains what ergonomics is, and why there is so much attention being paid to it these days, compared to only a few years ago.

Chapter 2 deals with the essential first step to health and happiness in the work environment: a self-assessment of your own physical and emotional condition. We suggest that you go out and spend a couple of bucks on a small, hard-cover notebook. As you conduct your self-assessment, write down the various elements. We'll show you how.

It is not our intention to make you feel depressed or morbid by noting everything that's wrong with you. On the contrary, not only do we hope that there's a lot *right* with you, we also want your personal record to indicate steps you can take at little or no cost to make yourself happier and healthier, and, above all, to make you more contented at work.

Did you ever stop to consider how much of your life is spent at work? We bet you have! But in case you've forgotten, there are approximately 8,760 hours in a year. Let's say you sleep eight hours a night; if we deduct those hours, we get 5,840 hours of waking time. Let's knock off 588 hours (3.5 weeks) for vacations and holidays; that leaves 5,252 hours. If you work only eight hours a day and never come in on a weekend, that would amount to 1,940 hours spent on the job. So you're spending 37 percent of all the time you're awake, and not on holiday, at the office! This is a rather conservative percentage; who among us has never stayed late, or come in to help out when we should have been elsewhere?

Furthermore, these calculations don't take into account the commute. Let's say that it only takes an additional hour a work-day to make the round trip. That's roughly 240 additional hours, and so you could really say that you spend almost 42 percent of your nonvacation time at work! That's heck of a commitment of time and energy. Given all the normal work demands made on you during the day, do you want this experience also to be a negative one for your body? Of course not!

We encourage you, therefore, to be frank in your assessment of your starting point: Be aware of your present pains, weak-nesses, or disabilities; your past physical and mental history; the medications you're taking; and how you treat your body (exer-cise, food, alcohol, smoking). We're not into giving lectures, but we do feel that starting by being honest in your self-assessment is essential for understanding your personal ergonomic situation.

Chapter 3 introduces you to your office environment. We think it can be described as a "pretty unusual place." You'll start to understand why as we guide you through the place you thought you knew so well!

For one thing, we're going to reintroduce you to your chair—the single most important item in your office, we believe, insofar as maintaining good health is concerned.

And because we don't want to be accused of favoritism, we're going to give time to the desk, the monitor, and other pieces of equipment, as well.

You'll be introduced to people who have been injuring themselves through their interaction with such items, and you'll learn how to prevent yourself from taking the same path.

Chapter 4 introduces you to the bones and soft tissues of the body that are susceptible to injury. We explain how injury comes about, even where there isn't any single incident that one can point to as "the" cause. We'll consider the neck, the upper back and shoulders, the forearms and elbows, the wrists, and the hands.

Chapter 5, "The Eyes," gets into the dangers presented by the intensive use of the monitor in your work; we tell you a little bit (not too much!) about how monitors work, so that you can see why they have to be treated with caution.

This chapter also touches on office lighting and sets forth some strategies for protecting your eyesight.

Chapter 6 gives you some easy, but highly effective, ways to avoid RSI. And since you should be both healthy and happy on the job, this chapter looks at some of the effects emotional stress can have on your physical well-being.

Chapter 7, "What Organizations Can Do," takes a look at three different organizations that are helping their employees fight RSI.

Chapter 8 is our attempt to predict the future of the work environment. For many centuries a whole segment of the economy was called "cottage industry." We seem to be headed back in that direction. For example, a greater percentage of us is going to be working, at least part time, out of the home, whether for ourselves as entrepreneurs or for companies and other organizations through "telecommuting."

In the centralized workplace, the emphasis will be more and more on the workgroup. A large percentage of us will be "connected," in the sense that we will be working with colleagues via a computer network. So wherever we may be, we'll be spending more time in front of a monitor, this time enhanced by video, so that we can take part in "real-time" meetings with our colleagues in the next cubicle, the next building, the next state, or anywhere in the world—perhaps eventually on the next planet.

Taking a look at what the future probably holds is an excellent way to emphasize the need for you to take responsibility today, and every day of your working life, for your physical and emotional well-being. There's a lot you can control, and much more that you can affect, if you have the knowledge. And to provide you with this knowledge is why we wrote this book. We hope you'll help spread the message to your family and friends.

Chapter 9, "If You Think You Have RSI," tells you how to go about getting help when you have a problem related to your work environment. It covers everything from how to identify the right people to inform about workplace hazards to what sorts of people provide appropriate health services. And we've emphasized once again the key role you yourself play in identifying problems early on and getting effective help *now!*

We hope you enjoy reading *ErgoWise!*

W.S.
R.C.

Acknowledgments

The authors would like to thank all those who helped in the creation of *ErgoWise*.

First and foremost, we thank MacTemps, Inc., the world's leading staffing company for Macintosh computer experts. MacTemps employs thousands of computer operators every day, and the company's concern for the long-term well-being of its people was the original inspiration for this book. MacTemps brought the authors together, and entirely funded all aspects of the creation and production of *ErgoWise*. Very special thanks goes to Jae-Ho Synn, Vice President, and all the wonderful and friendly staff of MacTemps' corporate headquarters in Cambridge, Massachusetts, without whose enthusiasm and support this project could not have been possible.

The staff at the Natick Research, Development and Engineering Center, U.S. Army Soldier Systems Command (Prov), was most supportive in setting up interviews with persons involved in ergonomics; our warm thanks to Kenneth R. Parham, Ph.D., Acting Chief, Human Factors and Ergonomics Branch, and to Claire Gordon, Ph.D., Senior Anthropologist, both of the Science and Advanced Technology Directorate, for granting us permission to visit their facility, and for arranging our interview schedule there. Dr. Carolyn Bensel, Steven Paquette, and Marcie Kronberg gave freely of their time, and we are most grateful to them.

Thanks, too, to Denise Gauley, RN, COHN, Manager of Medical Services, Phoenix Home Life Mutual Insurance Company, for her support of and participation in the combined medical and ergonomics endeavors at the Greenfield Clinic of Phoenix; to Jean Duprey, RN, nurse at the Greenfield Clinic, for her energetic and empathetic assistance in matters relating to patient care; and to Ian Bowers, vice president of human resources, Phoenix Home Life Mutual Insurance Company, for his essential support of Rab Cross's ergonomics projects and his contribution to Chapter 7.

Rab Cross also wishes to recognize the friendly cooperation of the employees of Phoenix in Greenfield in the process of interviewing them, studying their interactions with computers and other equipment, and examining their injuries; this has contributed greatly to his understanding and appreciation of office-related RSI.

Our thanks also to Robert Delaney, O.D., Hampshire Optometric Associates, Shelburne Falls, Massachusetts, for his assistance with the part of *ErgoWise* that deals with the eyes.

Dennis Mattinson, Corporate Ergonomic Program Manager, and John Vaughn, Certified Ergonomist, Sun Microsystems, Inc., Mountain View, California, found time in their busy work schedules to be interviewed on what this leading high-tech company is doing about the RSI problem, and our warm thanks go to them.

Sally Jackson, of Jackson & Co. in Boston, was an early supporter of the project, and we thank her for helping to get things under way. Thanks also to Brian Roeder, a summer intern at Jackson & Co., who assisted with the research.

Most especially we would like to thank our wives, Melinda Cross and Gesine Schaffer, and Rab's children Jamie and Megan, for their consideration and patience as we toiled at the keyboard and traveled across the country for work sessions and interviews.

ErgoWise

Chapter 1

Looking Out for Number One

Strange Places, Strange Doings

This book is a guide to the place where you work every day. Of course, your office, cubicle, workstation, or other workspace is familiar to you; after all, you spend enough time there each week! But what you may not appreciate fully are the very real risks of physical injury that you may be inviting while you are doing your work. It is increasingly evident that these risks arise from the way you interact with your work environment—more specifically, the way you use the furniture and equipment provided by your employer. Of course, the risks are also related to the design of the equipment.

What kinds of injuries are we talking about? They're mostly injuries to the upper part of the body and to the lower back. But they don't come about suddenly, as they might from a fall, for example. Instead, these injuries develop over a long period of time, usually without your knowledge. They're the result of the buildup over time—sometimes over years—of the effects of repetitive stress on the soft tissues of the body. That's why they're known as *repetitive strain injuries*, or RSI.*

How serious are these injuries? Very. They can keep you from earning a living; they can render you unable to perform some of the simplest activities of daily living, such as opening a jar, picking up a telephone, turning a doorknob, or lifting up a bag of groceries—or a child.

*Another term is cumulative trauma disorders, or CTD.

How real is the danger that you might experience such injury? Well, in the last decade in the United States, RSI went from 18 percent of all occupational illnesses to 61 percent.* That's a tremendous increase! Figure 1-1 shows the growth in RSIs from 1982 through 1993.

But, of course, whether *you* get injured is another matter. We've written this book to try to give you the information you need to avoid or at least greatly lessen your exposure to injury. Ultimately, each of us is responsible for working in a manner that maintains our good health. This isn't to say that employers don't have the responsibility to provide a safe working environment; they do. We, however, have to learn safe work practices, just as we have to learn how to drive a car safely. And after we've learned what we *ought* to do, we have to follow through and do it!

After all, before a skier hits the slopes, she has generally spent lots of money on equipment—on clothing, skis and poles, goggles, and boots. Before a rock climber starts scaling cliffs, he has invested in ropes, carabiners, climbing shoes, pitons, and picks. And before a scuba diver lets himself over the side of the dive boat, he has amassed his own specialized equipment. Also, each participant in these demanding sports has usually undergone some kind of training for the activity. Let's face it: Only the

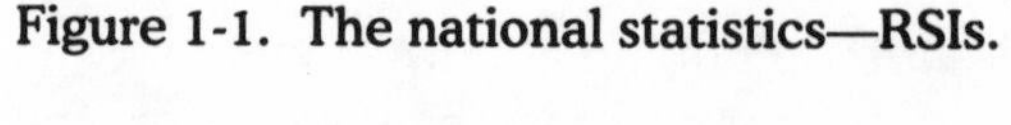

Figure 1-1. The national statistics—RSIs.

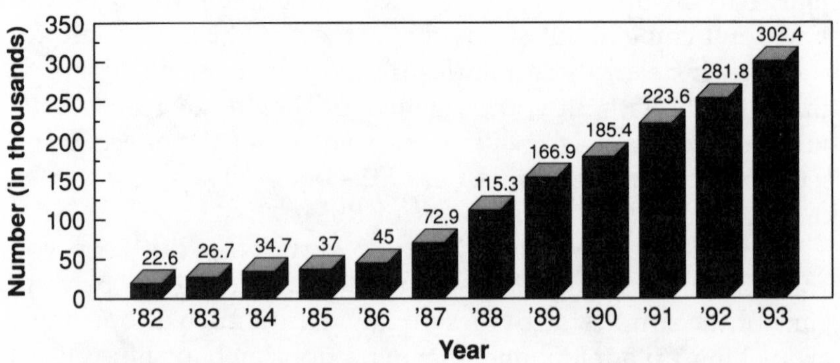

Source: **Federal Bureau of Labor Statistics.**

**San Jose Mercury News*, March 13, 1995.

most reckless among us would strap on scuba equipment and carry out a deep dive without getting instruction on how to use the equipment!

What happens when people interact with equipment without knowing its proper use? Some of us have known people who have tried the sport of alpine skiing and have let themselves be talked into going up to the top of the mountain before they've spent any time in ski school. Often we remember them because of the accidents that they've had, and the injuries—sometimes severe—that they've sustained.

Many environments, and many types of activities, are not natural ones for the human body. When we indulge in them, they place us at risk to some degree—risk of bodily injury, or even death. It's only smart that we take whatever steps we can to lessen the risk, so that we can enjoy the activity—and continue to enjoy it.

The modern office is not a natural environment for humans. Our bodies were not designed to sit motionless for hours at a time, with our fingers moving rapidly over tiny keys. Our eyes were not designed to stare at illuminated screens for hours; it's been discovered, for example, that when we look at a computer screen, we blink our eyes about two-thirds less than we do when carrying on a conversation with someone (seven times a minute versus a "normal" twenty-two times a minute). In a way, mountain climbing, scuba diving, and skiing are a lot closer to what is natural for us than our working life is, because they all demand a variety of movements of our bodies' large muscle groups, stretching, and constantly exercising our eye muscles by refocusing our eyes.

The Office Environment

One of the difficulties of modern life is that most of us don't really understand the office environment. Nor do we understand very well the equipment we use in this environment. Sure, we know how to use the desk, the phone, and the computer; we work with them all the time. But, in a sense, though it may sound odd, many of us do not really know how to use these common items in the best way, because we've not been trained. We are provided with the operating manuals, but these usually do not address the physical considerations recommended for safe usage.

Not only are we often ignorant about the inert objects in our work environment, but most of us really have only a passing acquaintance with the most important piece of "equipment" we work with—our own bodies!

Companies are becoming increasingly aware of the importance of making the office workplace more comfortable and safe, in an effort to lessen the risks to employees. Office employees themselves can participate effectively in this effort, and the first step in the process is to understand both how our bodies were designed to function and the sorts of specialized demands the office places on them. Once we understand these things, we can take some responsibility for our well-being while we're at work. Let's face it: no one really knows as well as we ourselves do what we actually *do* while at work. It's just that most of us are on automatic pilot a good deal of the time, and so we are not actively conscious of what we are doing, how we are doing it, and how it affects us, mentally and physically speaking.

One of the major purposes of this book is to help you achieve this awareness. It's important that you do this. Not only can you increase your own well-being, but you can help others to avoid strains, stresses, and injuries. You can also bring to the attention of the appropriate people various ways in which working conditions can be improved. Regardless of where employers may stand on the altruism scale, there isn't a single one of them that doesn't know that a healthy workforce is more productive than an unhealthy one. But, obviously, the more you know about your work environment and habits, and your body's reaction to them, the more cogent your description of how to make things better will be.

Ergonomics

This is probably a good point at which to introduce you to the term *ergonomics*. Ergonomics is a term dating back to the late

1940s. It describes "the science of adapting work and working conditions to suit the worker, rather than forcing the workers to adapt to the design of the machine."* One of the most interesting things about the science of ergonomics is that it is very much an individual science. If everybody were the same height and the same weight, and if their arms were the same length and their hands the same size, the answers to ergonomic problems would be much easier to find. But, of course, this is not the case. People have so many different dimensions that the combinations are practically endless. So ergonomics can be seen as a sort of complex puzzle—and, frankly, one to which we don't yet have all the pieces.

So why don't we just shrug our shoulders and give up trying to solve the puzzle? Because, as we've already said, workplace-related injuries are becoming more and more prevalent.

Older people in the workforce and retired people often comment that many of the work-related physical (and mental) problems people complain about these days didn't exist twenty or thirty years ago, or if they did, people just shrugged them off and got on with the job. Why the seemingly sudden focus on this sort of work-related ailment? For one thing, OSHA and the Bureau of Labor Statistics now follow reported workplace data, a policy begun as late as the 1970s. Also, in earlier years musculoskeletal injuries were regarded as part of life and aging, not the result of a job.

However, perhaps the most important reason attention is now being given to the effect of work on our bodies is the veritable explosion of health care costs in our modern society. High technology has moved into the medical field, giving us fantastic capabilities to diagnose and cure conditions that were considered hopeless just a couple of decades ago, or even just a couple of years ago. That's the good news. The other side of the coin is the very high cost—in equipment, in training, and in hiring of specialized personnel—of these technologies. Employers are seeking any way they can to reduce total medical costs, which—apart from salaries—constitute the single biggest expense in

*Frank Swoboda, "U.S. Seeks to Lessen Repetitive Injuries," Washington Post Service, Nov. 11, 1994.

American business. And one of the objects of their attention is the workplace itself.

There is another reason for all this attention to ergonomics, one we've already mentioned. A healthy workforce is happier and more productive. And increasing the productivity of employees, even by a seemingly tiny degree, can make a huge difference in being competitive in world markets.

Those of us who work in offices take no less pride than others in our ability to do good work and to work at a good pace. Physical and logistical impediments in the workplace can, insidiously, make us less effective, less productive, and less accurate in our work. And this can be extremely frustrating.

Yet another reason for the visibility of workplace-related ailments is the immense expansion in the use of computers and associated equipment in the office. The time will come when every productive person in America—from farmers to truckdrivers, from nurses to postal workers—will have a tiny piece of equipment in a purse or suit pocket, or dangling from a cord, that will combine the functions of telephone, computer, and television, and will vastly extend his or her communications potential. (Yes, we will be able to turn these things off!) At present, only a few of us have early, crude versions of such equipment, but most office workers already have personal computers or workstations on their desks,* most of them with a mouse as well as a keyboard.

The long hours many of us spend interacting with our computers affect us in many ways, not all of which are necessarily positive. Employers and employees alike are searching for ways to improve our relationship with the computer. Certainly, the availability in computer programs of every tool needed to analyze data, produce graphs and charts, and turn out faultless documents the first time means that we tend to move about the office less; we don't have to search for the dictionary or the whiteout bottle, go to the fax machine, and so forth. Everything we need is within reach, on-line, and so we tend to stay longer in the same place—in front of the monitor, keyboard, and mouse. Gone is the physical variety in our work tasks.

*In 1983, about 25 percent of the workforce used computers at work; in 1993, this figure had climbed to 47 percent.

Finally, office employees are becoming better trained and educated, and more independent of management. This independence is itself an outgrowth of the technology revolution, which has greatly boosted the productivity of individual workers. Management increasingly regards members of the workforce as partners in achieving the overall goals of the enterprise. When the business or government department has invested time and money in training, and when employees are expected to work more and more independently and to demonstrate creativity in understanding and meeting the needs of their department, there is strong motivation to provide the sort of work environment that will keep these partners in good shape!

Chapter 2
Self-Assessment

What Kind of Person Am I?

The topic of self-assessment will probably prove to be the most challenging one raised in this book—and it's very important, so don't skip ahead! In asking you to carry out a self-assessment, we're asking you to become very aware of your body type, with respect to several categories, and also of the accumulated physical wear and tear your body has experienced over the years.

Your worksite ought to match your body type as closely as possible. Unless you're into dictionaries in a big way, you may not be aware that there's a whole field of science called *anthropometry,* which wouldn't exist if there weren't variations—lots of them—in human proportions. Anthropometry (pronounced an-throw-POM-e-tree) has attempted over the years to derive such things as percentage distribution formulas that would show how many of us are of a certain height, how many of us have a certain waist size, and so forth.

A big customer for this sort of information has been the U.S. Department of Defense, which uses it, for example, to decide how many of what size boots to order for the men and women in the army. Your chances as a soldier of getting boots that don't fit are pretty low; there are, believe it or not, 133 different size combinations available to recruits, either on post or in a few days from depots. The army recognizes that properly fitting boots are essential to a soldier's well-being. Being able to keep on the go can even save a soldier's life.

We don't want to sound too melodramatic about it, but *all* the equipment and furniture you work with every day, while not lifesaving, can have a strong effect on your quality of life. We deal with this rather extensively in Chapter 3. But suffice it to say that it is quite rare to find a company taking the same approach to "fitting" your office furniture to you that the Pentagon takes toward clothing and equipment for its troops.

Ergonomics isn't an exact science yet, but it's very clear that even a rather small mismatch between, say, the length of your legs and the chair you're sitting on can have a significant impact on your comfort, productivity, and long-term health. Though each of us has a general idea of what size we are, most of us don't know, for example, the *exact* length of our legs from the inside of the knee* to the bottom of the foot, or of our arms from the armpit to the elbow. You'll be relieved to know that we don't want you to run around measuring yourself, or your colleagues. But we *do* want you to note where your measurements are going to cause you problems. And if you ever get directly involved in setting specifications for the office furniture you'll be using, you will, indeed, wish to have a feel for your measurements, simply to be sure that whatever furniture is proposed for you is going to *fit* you!

There are some clues that will help you determine whether you've got a special kind of "fit" problem. For example, what drives you crazy when you try to get the right fit in a clothing store? Is it finding the right shirt-sleeve length? Or the right inseam? To take account of variations in human body measurements, the clothing industry has developed specialized niches. Do you have to go to a "big and tall" man's shop? or a "petite" or "full-figured" clothing store? You can probably find something close to what you need. But even if you can pull something right off the rack, chances are you'll want to have some kind of adjustment made to get a "perfect" fit. We all want to be comfortable in our clothing, and to look our best; after all, unless we're on the beach or are confirmed nudists, we keep our clothes on for sixteen hours a day—more, if we're workaholics.

But just think: We spend eight hours a day at a minimum "encloaked" in our workspace. Why can't we strive to get the same fit there that we demand for our clothing? We can, and we should—but it's not all that easy to tailor your office.

*The anatomical term for this is the *popliteal fossa*.

Social pressures have succeeded nicely in making us avoid even thinking about some of this information, like the shape and size of our buttocks, unless it's with dismay or reservation. For example, we're certainly not ever going to subject ourselves to ridicule by raising a buttocks-related workplace issue with our supervisor—that guy or gal with well-shaped buns of steel and a ready snicker. Another illustration: A few years ago there was a particularly stupid song on the airwaves with the encouraging message that "short people got no right to live." One might say the same about those who write these sorts of songs. But let's face it: Though we know it's wrong, and psychologically destructive, to do so, most of us still believe that if we don't have the perfect shape (as defined by Madison Avenue and the ads for body building and salons), we've failed in some basic way.

Well, this book can't solve *that* problem for you, but it can tell you that lots of work, of excellent quality and high value, is done every day in this country by workers of all heights, shapes, and sizes. And each one of them should be doing that work in an ergonomically supportive environment.

What Condition Is My Body in?

Something to be aware of is that, just like the car you drive, your body has acquired some "dings" over the years. And given the way we live our lives in the late twentieth century, this is going to be true whether you're twenty, thirty, or sixty years old. Partly because of our attachment to the automobile, partly from our emphasis on sports at all levels of the educational system (sometimes to an extreme degree), and partly just from engaging in normal activities—say, picking up a crying child—most of us have experienced physical injuries to some degree. After certain of these injuries, our bodies don't completely return to normal—in fact, where significant muscle or tissue injury has occurred, changes continue to evolve *after* the "healing process" has ended, that is, when we've ceased to feel the acute pain associated with the original injury. These changes are often related to scarring of body tissues. We don't see these scars, the way we would if we had cut or burned ourselves, because they're actually inside our bodies, under the skin. But they are there.

Complete healing of certain injured body tissues—tendons, ligaments, and bones—can take as much as 120 days—four months. Very seldom do we allow our bodies sufficient time to recover 100 percent from such injuries. Football players in high school and college, as well as on the professional teams, take pride in "playing hurt." Tennis, skiing, body surfing, ice skating, and many other sports can result in injuries to muscles and tissues. Whether because of external pressures or from our desire to get back on the slopes or the field or the courts, we usually start in again too soon. And it's not just sports that cause these kinds of injuries. Confirmed couch potatoes can get up to get a snack, trip over Junior's robot toy, and fall, injuring a wrist or an elbow. Even resolute sluggards occasionally step off curbs or take the stairs (usually down). Curbs can cause ankle, wrist, and knee injuries, and the staircase is one of the most dangerous places in the home.

When we don't get signals from our bodies in the form of severe pain, we assume that everything is all right, when it isn't. These injuries may well have altered the way the body functions, at least in the short run—and perhaps permanently.

It is normal, after an injury has "healed"—even when we're aware that we've had some impairment or change in the way we swing our arms, or how much we can bend the right knee—for us to shrug our shoulders (if we can), accustom ourselves to the new situation, and get on with life.

What's really happened here is that we've redefined ourselves. If someone asks us how we feel after our injury, many, perhaps most, of us say, "Very well, thanks," because we're doing more or less OK, especially compared to people we know with severe and obvious physical problems. In any case, we have no choice, and so we smile and make the best of it.

This attitude may be praiseworthy in a psychological sense (no one likes a whiner), but it doesn't promote an awareness of physical issues that may require some additional attention in the work environment so that we can function better and avoid aggravating old injuries!

So we suggest that you make a little list of past injuries

you're aware of, and just keep it in mind as you read the rest of this book. It will be helpful.

We've gone on about this at some length because scarring, as well as the aftereffects of bone injuries, can result in reduced flexibility and scope of movement in the workplace years later! The funny thing is, because these changes occur over time, we are sometimes not even aware of them, or how severely they've affected how we feel at the end of a long working day.

The following measurements are critical for fitting your work environment to you: seated buttocks to eye height; leg length from knee to buttock and hip to floor; leg from inside knee to floor; arm from shoulder to elbow, and from shoulder to tip of index finger; arm from elbow to tip of index finger; width of shoulders (better get someone to help you with this one!); hips; finger span of hands; and length of fingers.

You'll see why these are important in a little while; right now we just want you to be aware of them.

What's Happened to Me?

We want to show you five areas commonly subjected to injury, discuss briefly why they are important, and then ask you to summarize your own past history of injury in your notebook. It's possible, of course, that you have never received any injury, although it's more likely that you may not recall an incident that really happened, particularly if it occurred long ago.

The Neck

For several reasons, the neck is one of the most important parts of the body from the point of view of well-being in the workplace. It holds your head up, for one thing. And your head is like a fifteen-pound bowling ball!

No, we're not trying to be cute here. We bet that everyone old enough to read this book has had some kind of neck pain at least once in her or his life. You know what it felt like, having to turn your whole body to look to the left or right or having to bring the paper up to your eye level, instead of just looking down.

The head supplies all the important nerve signals to your arms, hands, and fingers, all of which are, of course, critical to office work.* An injury to the neck puts all these functions at risk. The neck is particularly susceptible to injury in a car crash, of which there are plenty every year. Whiplash injuries are less prevalent now than they used to be in the days before head restraints were required equipment in all new vehicles, but there are still plenty of them. A whiplash can cause lots of damage: strained muscles, sprained ligaments, pinched nerves, and torn disk fibers. The aftermath of such an injury can be felt five or ten years after the event.

Whiplash and similar neck traumas can also result from sports activities, particularly football and other violent contact sports; they can also happen on the school playground in tumbles from slides or other play equipment. As indicated above, such injuries can result in a considerable decrease in head and shoulder mobility, which can, of course, affect comfort and performance on the job. In contrast, the long-term effects can be more subtle: perhaps pain in just one direction, or only at night. (Does that sound familiar to you?) That's why they should be taken into account in setting up your personal workspace.

The Lower Back

This is an area that tends to get hit hard by the activities of daily living. Lifting, pushing, and pulling the wrong way can all cause injury to the lower back. Is there an adult who hasn't at some time experienced a twinge of pain while lifting a bag of groceries from the trunk of a car, placing a child in a car seat, or, for that matter, taking the turkey out of the oven? Yard work, home repairs, and, again, sports can cause such injuries. And these are serious injuries, not because the back sometimes has to be operated on, but because the lower back is a critical part of the body

*Offices can be fitted for persons with severe impairments, such as paraplegics or quadraplegics; here we are talking about functioning in the usual office environment.

for most of what we do all day long at work, namely, *sit at a desk.* Whether we are staring at a monitor, writing memos, talking on the phone, or whatever, we are doing it while sitting. The lower back undergoes continual pressure from all this sitting; if it isn't in good shape to start with, it will get a lot worse without proper support.*

The Shoulders

Shoulders get injured in childhood and young adulthood from all sorts of activities, including throwing a ball. Dislocated shoulders are caused by contact sports, by falls, and by misuse of gym equipment. These sorts of shoulder injuries can result in the internal scarring we've already mentioned; the result can be limited shoulder motion, pain, and loss of strength.

We carry a lot of weight on our shoulders when we work, and not just in a figurative sense. Merely working at a keyboard or reaching for a mouse or the telephone puts stress on the shoulders; how much depends a lot on where the objects you use are positioned with respect to your arms. In a perfectly organized workspace, all these items would be easily accessible to you without your having to move your arms very much. Without going into a lot of fairly complex biophysics, let us point out that when you're using a keyboard positioned so that your upper arms can just hang down loosely from your shoulders, there's minimal stress on your shoulders. Less stress means less strain, and so at the end of the day, or week, you won't have any pain in the shoulder and upper back.

When we have to stretch out our arms to reach the phone, the keyboard, or the mouse, however, we exert pressure on our shoulders. And the greater the reach, the greater the force. The formula is: Force $=$ weight $\times$ distance. So it's the *reaching* that causes the problems.

All this may sound somewhat crazy to you. Don't we reach

*There are some lower back conditions that are nontraumatic; rather, they are hereditary. The most common is scoliosis. Scoliosis can affect seating comfort and, since it can make your body asymmetric, can result in a twisting or turning of the body that affects alignment with office equipment.

out for stuff every day of our lives? Sure we do. But remember, it's the *repetitive nature* of the action, time after time, day after day, that causes the kinds of injuries being considered here. Yes, you can also get shoulder stress from raking leaves, but this kind of pain tends to be short-lived, because you don't rake leaves eight or more hours a day, week after week, year after year (unless you're a professional gardener). And because you're often seated while you're in your office, the reaching you do is often at or above chest level—using the mouse, getting documents from an in/out basket, retrieving manuals, and so forth. The higher you reach, the farther you reach, and the more force is exerted on the shoulder.

This is why the setup of the equipment you work with is so important. It has to fit the unique individual that is *you*! For example, a heavy person is often forced by his or her physical shape to place the keyboard too far away to use it without reaching. Similarly, because of the upper body width, a heavy person is usually forced to squeeze the arms inward in order to get his or her hands onto the less-wide keyboard. Such a person's shoulders are subjected to a double strain: reaching and squeezing. Over the long term, this can lead to a painful situation. Though you'd think that the shoulders would be able to bear such stresses with ease, over time even a relatively small increment of pressure can exceed the strength of the muscles involved.

The Wrist

The wrist is another area that often bears scars from injuries. These injuries frequently come from sports, of course, but they can also come from lots of other activities. The growth in popularity of in-line roller skating has resulted in an increase in wrist injuries where the participant has neglected to buy (and use) wristguards. But a fall from *any* cause can cause injury to the wrist. You can roughly gauge the condition your wrists are in by seeing how they extend and angulate (see Figure 2-1). You should be able to flex your wrists up and down (with your fingers straight) almost 45 degrees in either direction. Likewise,

Figure 2-1. Flexing and angulation of the wrist.

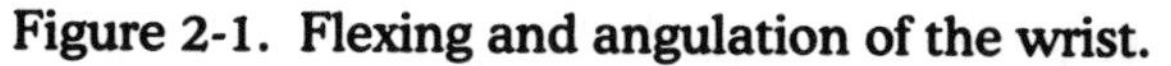

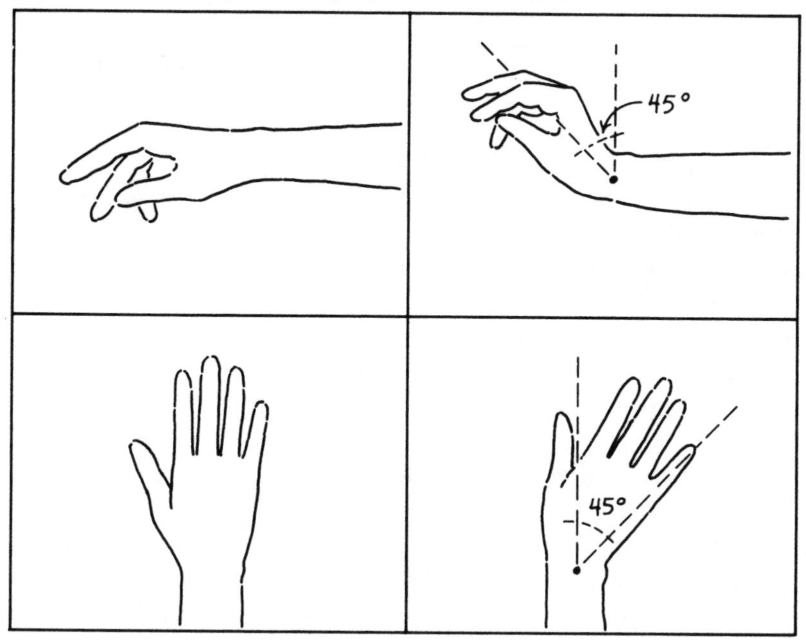

you should be able to move your hands to the left and right, with your forearms steady, about 45 degrees in either direction. If you can't do this, or if you experience pain in doing so, it is likely that you have had a wrist injury.

The wrist, as most of us realize, is extremely important in our daily lives. RSI can damage it to the point where you can't type, write, hold a book (or any other object), drive, or wave good-bye. Hand and wrist use is the key link between your brain and your keyboard, or your pen or pencil. Or, for that matter, between your brain and the telephone and many other things you use at work. Scarring of the wrist can result in reduction of the space in the carpal tunnel, through which pass the tendons and nerves that give you the ability to work with your hands. As the space gets smaller, pressure builds up on the nerves and can reduce the sensation in the fingers. This is an extremely serious situation, and surgery may be required to free up the nerves and tendons.

The Knees and Ankles

It's fairly obvious that we have to use our hands and wrists a lot in virtually any work we do. But why include knees and ankles in our self-assessment?

Like other injuries, damage to knees and ankles can result from a host of daily activities, as well as from various forms of sport. Injuries can, for example, start with tiny fractures brought on by running. You might not even be able to see them, but with repetitive stress, they can happen. In some cases the bone will actually change its shape. Or a tendon or ligament might become severely strained. Knee and ankle injuries are very common, and can be severe. Have you had old fractures, or torn ligaments or cartilage? Do you get a sore leg when sitting or driving? If so, how do you make yourself more comfortable?

When a person with such an injury is working in an office, what she or he does with the legs and feet can add to the stress, aggravating the injury. For example, a very short person may be working in an environment (chair, work surface) designed for a person of "normal" height. The short person's feet and legs may hang down from the edge of the chair, stretching the injured area. Or the person may sit in one of a number of positions—one leg doubled under, legs crossed, and so forth. These are, unfortunately, very common postures for shorter office workers.

Your Body's "Fit" to Your Workspace

In Chapter 3 we take you on a tour of your work environment. If you've written down approximate measurements and injuries to the above five areas of your body that you may have sustained, you should now be sensitized to your body type and to your history of injuries. It may well be that after you complete the tour, you'll realize that you've been trying to accommodate your body to your environment, rather than the other way around. If that's the case, it's time to make some changes.

Chapter 3

Understanding Your Workspace

Learning the Terrain

It's a place you go every workday. You've got a couple of miniature cacti in pots and photos of your kids or pets. On the sides of your computer monitor are some yellow Post-its with the new phone number of your dentist and a reminder to pick up a birthday card for your significant other on your way home.

This is your workspace. It's a place that holds no mystery for you, you know it so well. It may not be where you'd prefer to be if you had your druthers, but unless you win the lottery, it's where you're going to spend lots and lots of time.

But as we suggest in Chapter 1, you may not know this place as well as you ought to. And we believe that after reading this book, and particularly this chapter, you're going to look at it in a very different way. Our aim is to make you knowledgeable about ergonomics, at least to the degree that you can understand what is right and what is wrong and harmful about your work environment. Relax! It's not as difficult as you may suppose. And it is very important. We expect that many readers of this book will make some significant changes in their workspace—and will do so right away, not sometime next week. It's also likely, we feel, that hitherto unnoticed problems will be brought to the attention of facilities managers, or whoever has the responsibility for selecting office furniture and determining office layout.

We're going to start with the common pieces of office furniture you use every day: the chair, the desk, the computer (monitor, keyboard), and the files (desk drawers and cabinets). There's more to know about these innocent articles than you suspect!

Then we'll turn to the office layout—how these things are arranged with respect to the unique person that is you, and also with respect to lighting.

We hope you enjoy the trip!

The Chair

The chair is the most important piece of furniture in our workspace, from an ergonomic point of view, which means from the point of view of keeping us comfortable and functional. And it's the one we pay least attention to, probably because we don't actually see it most of the time we're at work. (How could we? We're sitting on it!) Its relationship to the stuff we *do* look at all day—the monitor screen, the desk, the notes we're working from, the telephone, the file drawer—is also important; we'll consider this when we get to the office layout.

Figure 3-1 shows drawings of several different kinds of chairs. Which kind do you think you would find in an office environment? If you're experienced in taking multiple-choice tests, you probably guessed—correctly—all of the above.

Among the reasons for there being such a variety of chairs in the workplace is that those in charge of spending money to buy them either may not be aware of what to look for in a chair or simply may not have—or may think they don't have—the money it takes to buy them. A really good chair purchased new at retail costs between $300 and $1,200. That may appear to be beyond the reach of many businesses.* So chairs tend to be purchased en masse at the lowest possible price—just something to keep employees' bottoms off the office floor.

Also, companies tend to purchase chairs (and other furniture) in a piecemeal fashion over time, especially if the company

*In fact, for businesses of a size that permits them to buy in bulk, i.e., at a decent discount, there's economic justification for buying chairs that are ergonomically sound.

Figure 3-1. Seven different kinds of chairs.

is growing. Often, too, the managers of a company get first shot at the new stuff. The old chairs get put into a company storage area, from which they may be redistributed all over the company. What ensues is a sort of shuffling process, so that it's not unusual to find lots of different types of chairs, even in the same department.

Some chairs used in offices clearly can't be adjusted at all to take account of the way you're constructed—for example, chair d in Figure 3-1.

Some chairs have just one or two adjustments that can be made. The height of the seat is the most usual one, followed by the height of the backrest. Normally the occupants of such chairs try to supplement the adjustments and add to the comfort factor by using pillows. Older chairs may be "frozen" in their present settings, making it a major project to try to increase or decrease the height of the seat by an inch or so. Some chairs, of course, do have multiple, easy-to-use adjustment features, and these chairs should be the easiest to adapt to a user's particular needs.

If you're at work, take a couple of moments now to check out the kind of chair you're using. How many kinds of adjustments can you make to adapt the chair to your body type? Do you know why being able to make them is important? Do you know how to make these adjustments? Unless you have a new chair, chances are that the operating instructions are missing. Can you make adjustments easily, without turning the chair upside down? Can you make them safely? And, *have* you ever made them?

Chairs and the Lower Back

Let's take a look now at one of the more important areas of the body, the lower back. It often causes us trouble—trouble that is often related directly to the way we sit, and hence to the chair. Figure 3-2 shows what we look like when you really get down to basics.

Note the area of the spinal column called the lumbar spine. It's this area that's affected when we complain of "lower back pain," a malady that afflicts many millions of us. Note that it's got a sort of reverse, or inward, curve to it. The drawing shows

Figure 3-2. Side view of the standing skeleton.

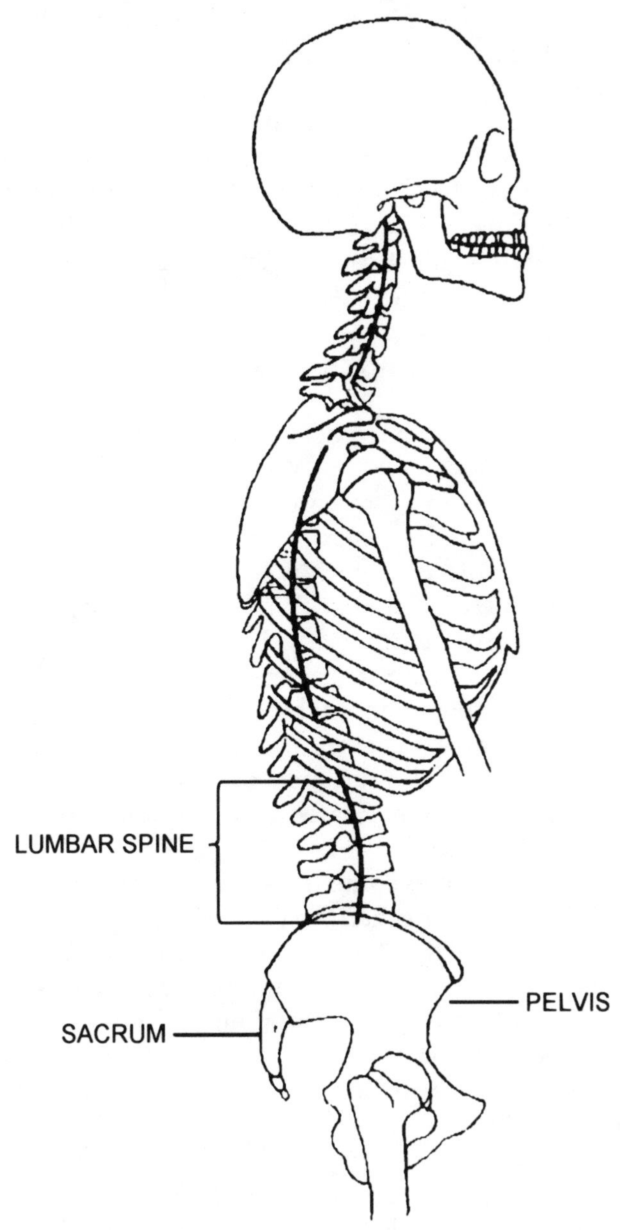

Source: Courtesy of illustrator Paul Schaffer

the normal curve of the lower spine; it's easily observable in a standing position. Aha—but when we sit! That's when the troubles can begin, because to the degree that the normal lumbar curve isn't maintained, the pressure on the lumbar disks is increased. You may already know that the body's main nerve channel, the spinal cord, along with its many branching nerves, is protected by these disks. When the disks are unduly compressed, the spine bones, the spinal ligaments, and the exiting nerves begin to grate against each other. The bones and nerves are very sensitive parts of the body. Statistics show that they are frequently injured as the years pass. Remember that in Chapter 2 we talked about all the daily activities that can cause strain on the lower back. Back injuries can be seriously aggravated by sitting the wrong way. And even a healthy back can get very painful in a bad chair.

There has been a lot of research in recent years on how much pressure is exerted on the lower back by different postures. The least pressure is exerted when we stand. Maybe this is why Leo Tolstoy wrote *War and Peace*—all 1,400 pages of it—standing up! Perhaps he instinctively realized that various seated postures and activities can increase disk pressure up to twice the pressure we experience when we're standing up.

An injured lower back is no fun, as too many of us already know. Low back pain and associated disorders affect eight in every ten adults in the United States at some point during their lifetime.

These conditions hurt—sometimes a lot. And sometimes we can't sit at all in any chair other than the one or two that we're "used to." The reason is that the other chairs increase disk pressure—perhaps only by a very small amount, but that's enough for the alarm bells to go off and the pain to come stealing in. People adopt all kinds of strategies to deal with this in the workplace, including, as we've already mentioned, the use of pillows and cushions. But this really doesn't tackle the crux of the problem, which is that the chairs these folks are using, which may be perfectly fine for somebody else, aren't right for them. In Figure 3-3, the small pillow, provided by a well-meaning therapist, sticks into the employee's back, and hence his upper torso is unsupported. (Also note that the armrests, nonadjustable, are

Figure 3-3. Using a pillow; this doesn't solve the fundamental problem, and can create other difficulties.

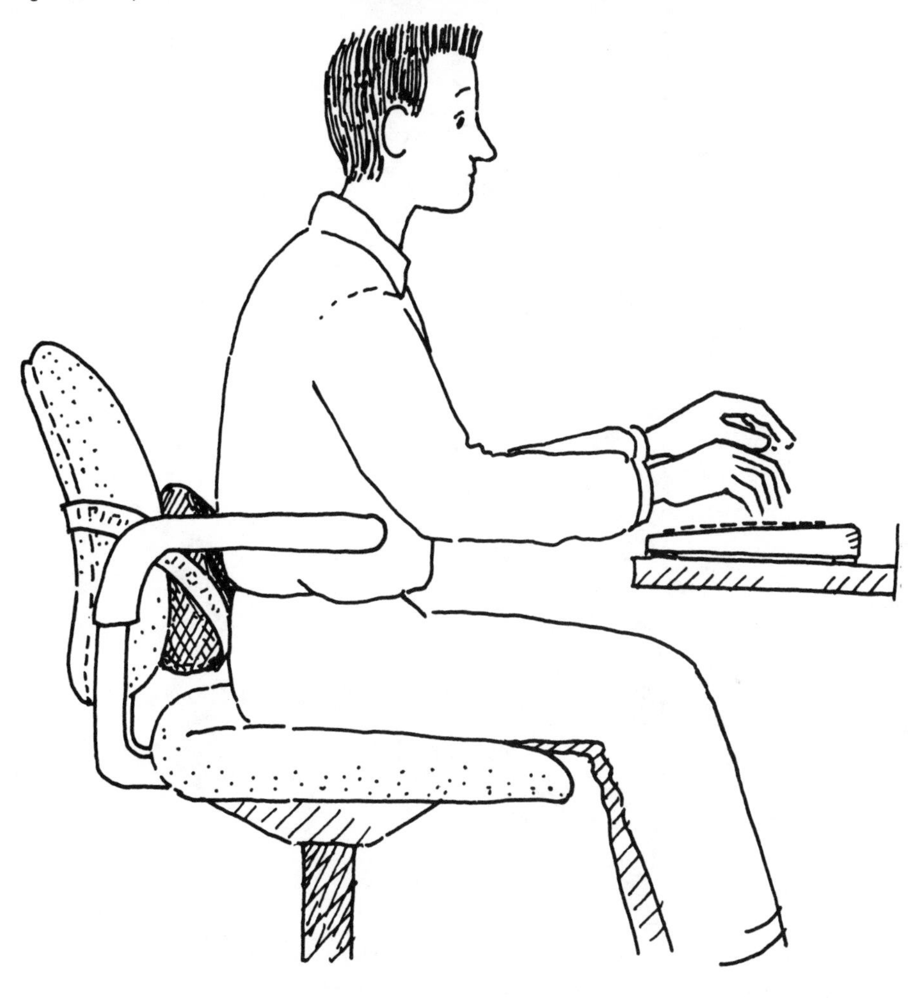

sticking into his elbows.) In the actual case from which this figure is taken, the user is a software developer, and he complained of not being able to concentrate. It's no wonder!

Figure 3-4 shows how one employee tried to adapt her chair by using three pillows. She informed us that she tried to avoid her workspace "as much as possible!" Chair d in Figure 3-1 was

Figure 3-4. Using three pillows in trying to adapt the chair.

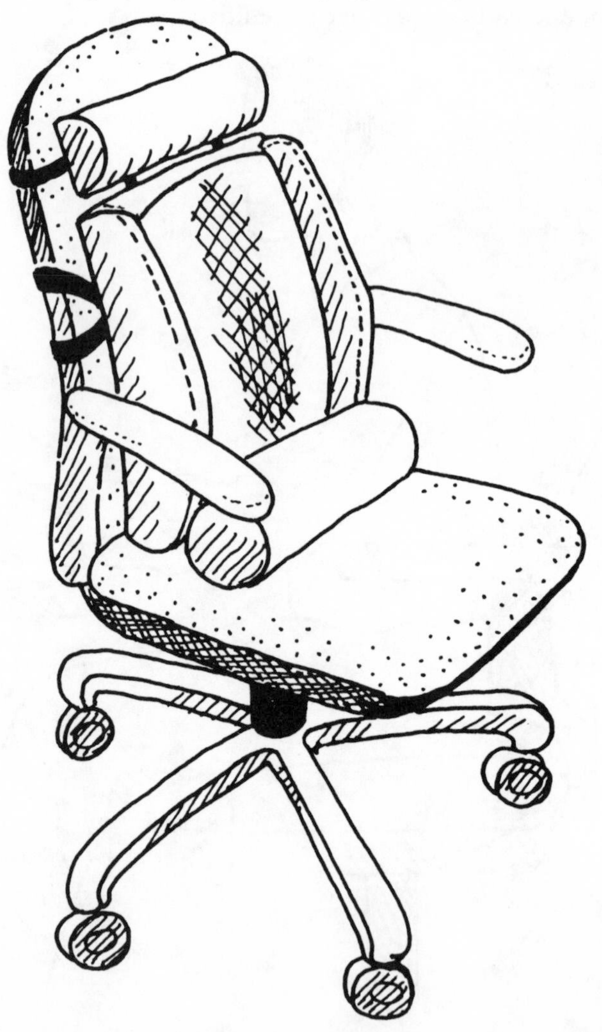

used by a young lawyer with both lumbar and cervical (neck) spinal disk problems, who experienced daily pain. She and the software developer eventually both found the same solution—a Zackback chair (see Figure 3-5), which is adjustable to their statures and back conditions. The young lawyer is matched quite well by her chair. You can see how the backrest fits into, and promotes, the natural lumbar curve. This woman now fre-

Figure 3-5. Employee matched quite well by her chair.

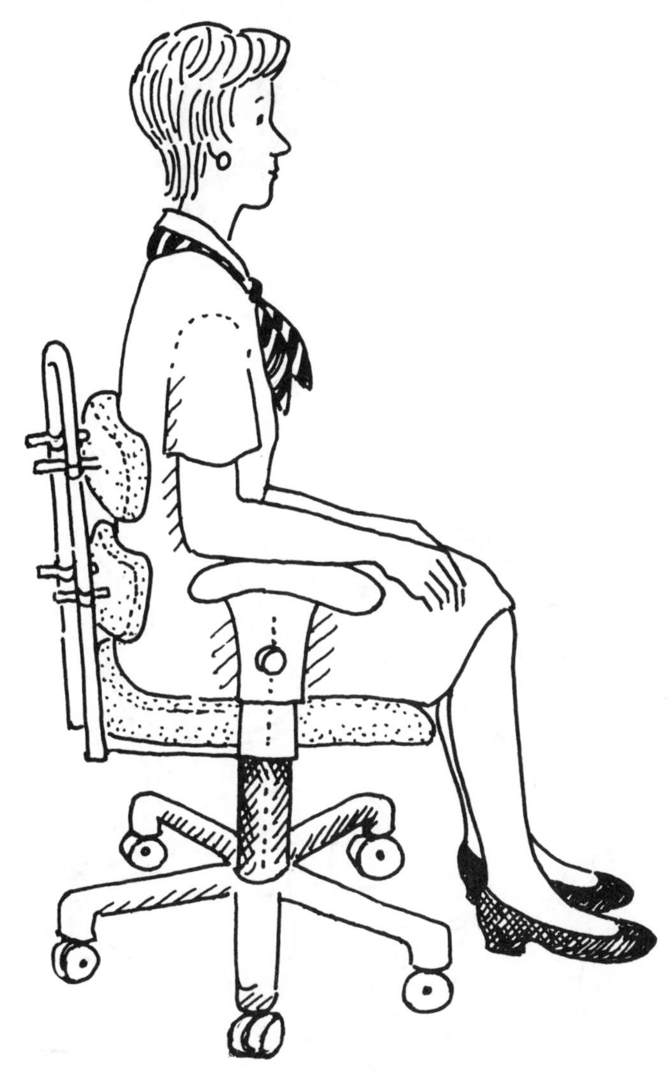

quently mentions to her fellow employees how pleased she is to have this chair for her job.

One of the nice things a properly fitted chair can do for the user is alleviate pressure on the lumbar disks. In Figure 3-6, the user is getting minimal support for his lower back. He's sitting straight, with the upper arms extended forward reaching for the

Figure 3-6. User receiving minimal lower-back support.

keyboard. Ideally, the upper arms should be in the neutral position, hanging straight down parallel to the torso. The pressure on his lumbar disks is about twice what it would be if he were standing. It's not that the chair is bad; it's that it is not right for *him*, because it doesn't support his lumbar spine, let alone his upper back.

Many offices still have chairs for which no adjustment is possible. Some of these look quite modern and comfortable, but of course they were designed for the "average" person. If there's one thing that can be said about the average person, it's that most of us don't resemble him or her very closely. Larger people who attempt to work in nonadjustable chairs have particularly serious problems. Not only are they denied lumbar support, but the backrest may press against the upper buttocks, accentuating the tendency of the buttocks to thrust the person forward and upward—not the most comfortable position to work in!

Oddly enough, some employees swear by older chairs, such as the one shown in Figure 3-7. This chair is virtually an antique, and is austere. It only has four legs (the recognized standard these days is five, to avoid tipping), yet there is plenty of space for the buttocks beneath the backrest. That means that, as ugly and uncomfortable as it may look, it's still a preferred piece of furniture for certain people with bigger bottoms.

It's not only office employees who may find themselves with the wrong kind of chair. Figure 3-8 and 3-9 show chairs reserved for managerial personnel at one company. Note that there is no contour for the lumbar area, and that the backrest isn't adjustable at all.

Chairs and Other Body Parts

We've been concentrating on the spine up to this point. But of course, there are other parts of the body that can be affected by the wrong kind or size of chair. If you are short, the "seatpan" of your work chair may be too long for you. That is, the length between the back of the seated buttock and the back of the knee is shorter than the length of the seatpan, and so the front edge of the chair exerts pressure on the popliteal fossa (remember this term from the footnote on page 10?). In case you think this

Figure 3-7. Older chair.

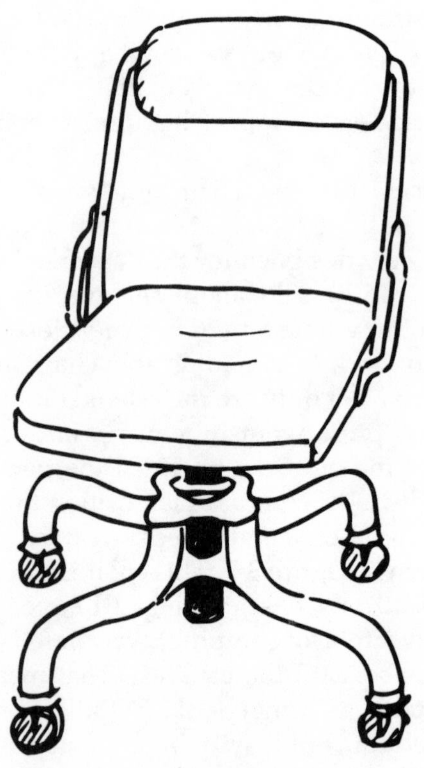

doesn't hurt, try wedging your fingers up against the back of the knee. Hit the right spot and you won't be able to stand it for more than a few seconds.

You may be starting to see how you can use the observations you made as part of your reading of Chapter 2 to judge the suitability of your office furniture. Maybe up to now you've thought of yourself as fairly tall. From the ergonomic point of view, however, "tall" people can be too short, so far as the seatpan of a chair is concerned; all their height might be in the lower part of the leg, or in the torso. And, of course, short people may have the reverse situation—their upper legs may be long enough to fit the seatpan of the chair, but their lower legs may end up dangling. This is because a shorter worker often has a lower leg that is not long enough to reach the floor. The result is that blood circulation is cut off, and nerves in the knees become sore. Figure

Figure 3-8. Executive chair.

3-10 shows an office worker who's 5′1″, with short lower legs, trying to solve these problems. He ends up sitting way forward. Result: no lumbar support—and when his arms are extended as in the illustration, the lumbar disk pressures are painfully increased. Figure 3-11 shows how the seatpan jams into the back of the knees of the person with short thighs.

Another "solution" to the short-legged employee's problem of being unable to reach the floor with his or her feet is to curl the legs up and sit on the feet. Or the employee may perch the feet on the legs of the chair (Figure 3-12), effectively removing

Figure 3-9. Executive chair.

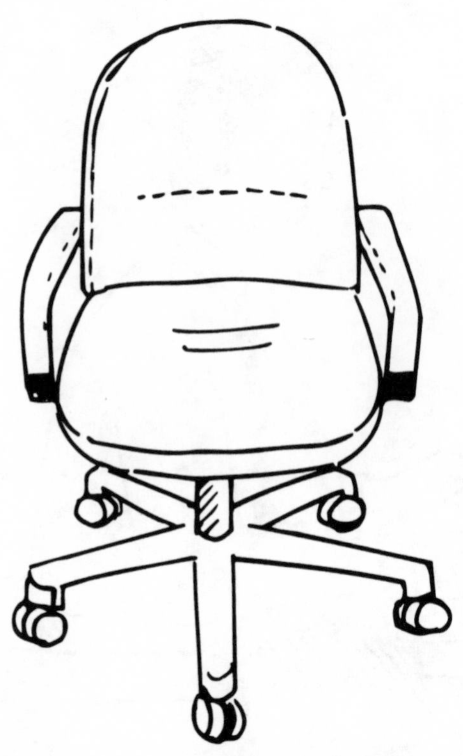

the lumbar support and putting the knee at an extreme angle. Sometimes an office worker will rest her or his feet on a shelf or box under the desk; every minute they stay there places a strain on the knees, fatiguing the knee joints, and reducing the lumbar support from the backrest. Even piles of books are sometimes used to prop up the feet. Unfortunately, they are neither comfortable nor very stable; having to bend down to restack them can present its own health hazard.

The Desk

The desk is just as important as the chair. And it's not just the piece of furniture itself that counts in determining how well your workplace is adjusted to you—it's also the arrangement of

Figure 3-10. Worker with short lower legs in too-high chair.

Figure 3-11. Worker with short thighs in overlarge seat.

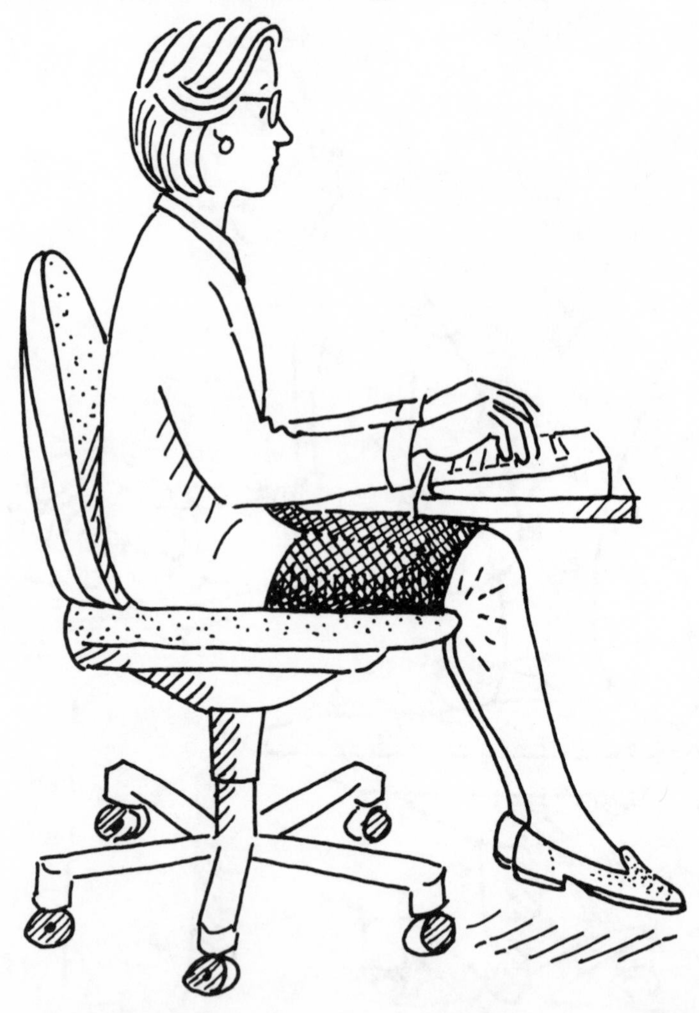

all the stuff *on* the desk, because you interact with this stuff all the time, every day. So first we'll tackle the desk as a physical object, and then we'll get to all the equipment and other material that's on it and in it.

Manufacturers of chairs are getting onto the ergonomics bandwagon, but most of the desks we see coming onto the market still seem to be in the same old format: four legs, a center drawer, side drawers (and hence a little sort of square tunnel for

Figure 3-12. Worker with short lower legs perching feet on chair legs.

your legs), and a modesty panel at the rear. The adjustments increasingly in evidence on chairs just don't seem to have made it to most desks.

Why is this? Well, for one thing, the ergonomic design movement is relatively new, and there's a lag time between perception of the need and the meeting of the need. But probably the main reason is that workers are much more conscious of the chair as an important part of their office life than of the desk.

After all, you put yourself in the chair; the desk is just there to hold all the junk! What this boils down to is that you, as the end user of desks, can help to create the market for better products. That's one reason why you're reading this book.

OK, so take a look at the desk you're now using. It may be an early version, of wood or metal—a heavy brute, with limited leg space. We've seen how chairs should be designed to take care of variations in leg lengths and body sizes. Should this same consideration apply to the legspace, or legwell, of your desk? You bet it should!

First of all, let's consider a large-bodied person. The legwell can be terribly constraining and very uncomfortable for such a person. Such a worker may be stuffed into this space, literally unable to move sideways even a few degrees. Fairly often, a keyboard or monitor or both is placed to one side or the other. In order to work directly facing these items (as is required for viewing the screen and typing), the obvious thing for you to do is to turn in your chair. If you cannot move from side to side, when you have to turn, you have to do so by twisting either your shoulders and upper back (a midback twist) or your lower back (a lumbar twist).

Figure 3-13 shows a woman who has been in physical therapy for a long time and suffers from midback pain because of the extreme twisting required by her workstation. Her upper back is deprived of any support (so that all the help given by the ergonomic chair is wasted). Her arms are now reaching in unequal fashion. And she would fall over to one side, except that her legs twist in reverse, or press firmly against the side of the legwell, to equalize the contorted physics of her seating. She has to twist this way because normally she sits facing straight ahead, so as to deal with patients coming to the service window. There are actually two legwells under her counter, but they're separated by a support structure, so she can't shift her legs to the left.

Since this woman isn't large-bodied, she doesn't twist very much, and so her muscle strain and pain are less. They can, of course, still affect her performance and cause her to miss hours, or even days, of work. But beyond that, how on earth can she, a devoted and skilled employee, give her full attention to the job

Figure 3-13. Woman twisting at her workstation.

when she might as well be in a coffin from her hips on down? *Comfortable* employees can work up to their potential. And that's what employers hired them for. So it doesn't make much sense to let an inert piece of furniture upset the equation.

Slim people can also suffer from the "coffin syndrome." In our research we ran across Ted, tall and skinny, who was trying to do a report with his knees crammed into a tiny legwell lined with cartons. He had no available legroom at all! Work had to be agony for him, yet he had been doing this for a few months, waiting for storage space for the cartons to open up somewhere.

Small stresses over time are precisely what give rise to office injuries.

A similar cramped leg problem can be caused by a center drawer. In one case, an office worker had to raise her chair so that her legs and feet would be in a comfortable position, but that forced the tops of her thighs to be jammed up against the center drawer. Talk about a no-win situation! Shortly after we observed this situation, the center drawer was removed, the desk was lowered a bit, and this particular problem was solved. In some cases, though, the desk can't be lowered; hence the keyboard ends up being too high. But this may be the best you can do . . . or is it?

In another case, a friendly coworker suggested the installation of a keyboard drawer, tray, or holder that would attach to the top of the legwell, thus bringing the keyboard lower.

This sounds great—but watch out for this sort of patchwork solution. Such holders often have shelves or brackets underneath them to hold the keyboard. These can cause damage to knees and clothing. Also, now you've got to reach over and beyond your keyboard to use the phone, calculator, or mouse. You've guessed it! This can cause a series of light, but repetitive, strains on the shoulder or back. Remember, we're addressing a problem known as repetitive strain syndrome; it can strike in many different body locations.

Finally, when you slide the keyboard holder back into the legwell for a session of deskwork, you still can't raise your chair because you've replaced the obstructing pencil drawer with an obstructing holder frame.

Not only does the tight fit cause a risk of injury; there is also an absence of ventilation, or air flow. Things can get pretty hot and sweaty as the day wears along. The way most employees cope with this is to shove the chair backwards, letting some air get to the lower body.

What this does is move the upper body farther away from the keyboard and mouse, which in turn causes strain and soreness in the arms, shoulders, or back. So the hapless worker moves the chair forward again.

What on earth is going on here? Simple! Human beings are trying to adapt themselves to the environment, instead of the other way around. For about eight hundred centuries, the story of civilization has been one of changing the environment to fit us; now the shoe is on the other foot, so to speak. Fortunately, there are some possible solutions, and later we'll see how companies can go about it. But an absolute key to any solution is *being aware of the problem.*

Newer desks (and some very old wooden ones) have wider legwells, and may be wider in general. Some units come without a pencil drawer, or have it to one side. Some desks have rounded edges or softened edges, so as to minimize pressure on the underside of the arms.

Replacing desks in some work environments are "work surfaces" that can be affixed to the paneling that forms the (usually movable) office or cubicle wall. Depending on the manufacturer and model, these work surfaces usually offer more legroom. Also, such work surfaces are generally adjustable (though often this takes a couple of experienced persons). Sometimes file drawers, which are constraints to movement in the traditional desk, can be replaced by storage boxes that are suspended above the work surface, or by rollaway files beneath the surface.

The Desktop

Let's take a look now at the top of the desk. Believe it or not, what's on it, and how these items are arranged, can affect your health. And what's on the desk is pretty different from just a few years ago—in case you hadn't noticed.

In the old days, one of the gifts you could always give Aunt Sally or Uncle Bill, when that promotion came through, was a classy "desk set," consisting of a big writing surface with blotting paper, with leather corners. If you really wanted to make a splash, you could also throw in a marble double pen holder with a built-in brass desk calendar. Your relative could come in to work every morning, twist a couple of little knobs, and know that it was Monday, April 22, for example.

These days, such a present might still be OK when Aunt Sally finally makes it to the presidency of the bank; she'll have a big wooden desk, and her computer will be on its own little table off to the side. For the rest of us, though, there's generally no special place for a computer, and if anyone should give us a fancy desk set, we wouldn't have room for it. Sometimes even the photos of our kids and the dog* have to go on the wall or the bookshelf. That's because the desk set has turned into a humongous computer (monitor and CPU†), or at least a large monitor and keyboard (if the CPU itself has been moved down onto the floor next to the desk, or you're only using a terminal).

In those earlier days, desks weren't designed to accommodate such large pieces of equipment. And even today, most of the desks that are in offices, even pretty modern ones, are modeled after the same old wooden desk, with the same old dimensions. At the same time, monitors have gotten even bigger. Let's see how that affects the placement of the monitor and, in turn, how that placement can affect you.

Computer monitors should be placed right in front of you; otherwise your neck, shoulders, and torso may find themselves doing the twist. But it's not easy to find the right fit between people and monitors. And monitors are difficult things to move around! If you need to look at some documents and try to push the monitor away from you, you may find that after moving it a few inches, it's back against the wall. Pull it forward too far and you also run out of room, and may find it in your lap. Also, of course, the monitor is wired to a power source and to the CPU; moving the monitor means moving this wiring. For some perverse reason, the wiring always seems to be just a bit too short.

The problems associated with the position of the monitor very often involve the head, neck, and eyes. In keeping with the plan of this book, which is to show you lots of the *wrong* elements of the workplace before we start suggesting what to do about it, we'd like to introduce you to some more positions taken by employees who find themselves in trouble.

Many people with bi- or trifocal glasses have problems with the vertical placement of monitors. If you wear such glasses, pay

*Sure, sometimes we also have room for the photo of our spouse or significant other.

†Central processing unit, or the "computer box" where all the action takes place.

particular attention to the placement of the monitor, or investigate getting special computer glasses.

The person depicted in Figure 3-14 is fairly comfortable except for suffering a bit of shoulder and neck stiffness and pain. The reason is that her monitor is just a bit too high (even 3 to 4 inches makes a difference), and so she has to cock her head back—not much, it's true, but enough to cause difficulties. A more exaggerated posture, but not untypical, occurs with a very high monitor. The telltale sign is when someone leads with her chin. This is a position to avoid! It places stress on the disk and nerves in the neck and can lead to severe injury over time.

It seems as if lowering the monitor would be a simple solution, but it isn't. Sometimes the monitor is on top of the computer, or CPU, itself, which means that it can begin at 6 or 8 inches above the desk surface. Now the issue may become where to put the CPU, and some employees are afraid to touch the darn thing.

Figure 3-14. Using a monitor that is too high.

We're happy to report that with most desktop computers you can put the CPU wherever you want—even in the next room, so long as the cables between it and the monitor are long enough—and it won't affect performance at all. And you can stand it on its end or side, so long as you don't block any ventilation openings or any ports, which are the places where you plug things in.

In some offices the monitor is placed on a special adjustable "boom" arm. But these things will allow the monitor to go only so low and no lower. If that's not enough, up goes the chin. Of course, whether you're short or tall affects the viewing angle. The next time you sit at your desk, take notice of your head position. Are you looking upward when you look at the monitor? If so, then something needs to be done.

Additionally, looking too far downward at the monitor can also be a problem. This can especially be a problem with laptops, except for those where the monitor can be detached from the rest of the system. Take a look at the viewing angle used by the man in Figure 3-15. This posture may be suitable for a (brief)

Figure 3-15. Looking down at laptop computer.

period of meditation, but not for preparing that fifty-slide presentation justifying locating the new plant in Hawaii. His upper back is unsupported, his shoulders are rounded forward, and his neck is slightly strained as well. A nice fat dictionary under the laptop could solve his neck and shoulder problem, at least until he has to look up a word.* But then, of course, as he uses the keyboard, his arms will go up! If he raises the laptop to correct viewing height, he should use a separate keyboard, properly positioned so that his wrists and arms will be properly placed; otherwise he's going to run into trouble. He should also use a separate mouse.

You've looked at the vertical and horizontal placement of the monitor, and wisely decided that it ought to be in front of you, at just the right height. But now you've got another problem. Often, when you use the monitor, you've got to refer to some document or other—a report, publication, price list, shipping schedule, or whatever. The question is where to put the darn thing. At one of the companies we visited, we found a young woman doing some writing while constantly referring to a large manual—too large for her document holder. So she had put it over to her left on the desk. Not only did this strain her shoulders and upper back, but the focal length required for her to read the more distant upper lines of the flat manual was greater than the focal length of her eyes. Attempting to compensate, she would squint, and strain her eyes.

The man in Figure 3-16 has the same problem (a common situation), the computer and keyboard occupying all the real estate in front of him. His head is bent forward and twisted to the left. A document holder can solve the bending problem, and basically solve the twisting one, so long as the holder positions the reading material next to the edge of the monitor.

Even where there's space in front of the employee, there may be risk to the neck. In Figure 3-17, see how the placement of the material this woman is working from has forced her to lean forward (no upper back support) and bend both her upper back and her neck. Not recommended! Figure 3-18 shows how dramatically a simple document holder has fixed the situation.

*If he is like most of us, he won't look it up, anyway. He'll ask someone.

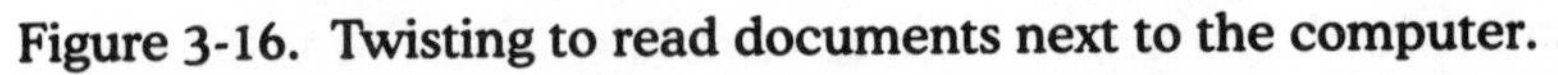

Figure 3-16. Twisting to read documents next to the computer.

If you commonly work with large volumes or thick, bound files, you can try using a cookbook stand to support the material. New models of document holders that can accommodate larger texts are now on the market.

Even with a well-placed monitor and with the document in a holder, things can get pretty twisted when you do simple things such as taking notes. In Figure 3-19, a worker is trying to read a message she's received over her company's network; at the same time, she's writing some notes. She's had to twist her torso to the right, reach out with her right arm to write, and twist her neck and head in the other direction in order to see the monitor. Her setup is really better for a left-handed person. Probably the reason she's doing the writing on the right side of the desk is that that's where the phone is located. It, and her writing space, should be switched to the left.

High monitors can cause problems to areas other than the neck. Have you ever run out of space on your desk, and just *had*

Figure 3-17. Leaning over to work from document.

Figure 3-18. Using document holder.

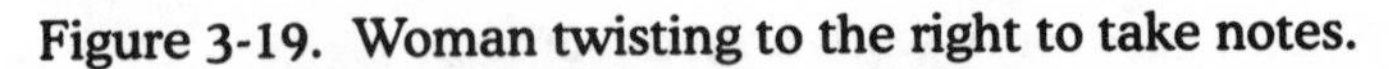

Figure 3-19. Woman twisting to the right to take notes.

to use the top of your monitor as a little storage area? Lots of us do just that—finding a creative solution to the storage-space problem. But it's better to have a storage-space problem than the shoulder and arm pain you may suffer if you have to reach up to the top of the monitor many times an hour. If you were asked to do this motion with half-pound weights at the health club, you would probably stop the motion as soon as the pain appeared. But in the workplace it may take many days or weeks before the pain caused by such "reaching strain" is first noticed. Most employees have never associated such pain with what they were doing to cause it. They don't suspect that such light reaching could cause a problem.

The Telephone

We've already mentioned the phone, so now let's see how this little monster can cause us harm—aside from the fact that it's always either going off and wanting us to answer it, or forever reaching a voicemail announcement instead of the person we're trying to call.

As you might suspect, it's the part of the phone we pick up that causes the neck and shoulder problems. And that's usually because in today's frenetic office (and home) environment, we're usually trying to do at least one other thing while speaking on the phone. In Figure 3-20, a customer service representative is taking notes as she deals with a customer call. Her right shoulder is elevated, and she has also flexed her neck to the right side, to hold the phone. Almost all of us know this position. If she does it constantly, day after day, as this woman does, she'd better switch right away to a headset with attached microphone, or she'll soon be seeing a massage therapist or chiropractor—at best. In fact, she's probably seeing one already.

Many people who must spend a lot of time on the telephone use a receiver mount as a partial solution to the problem. But even with this item, the shoulder and head are still bent toward each other to hold the mount. The stress is exacerbated when other pieces of equipment are poorly placed; we've seen a telesales person with her computer monitor to one side, so that she had to twist as she typed. This placed additional strain on her upper body. People who work at home run into the same problem with portable phones. They're very convenient, but there's just no good way to hold them other than by using one's hand.

Even when you use the receiver properly, other parts of the body can get into trouble, although this time the phone is an innocent bystander. The problem arises when someone who uses the phone a lot also has to make entries into a computer. The woman shown in Figure 3-21 has located her computer on the far right edge of her desk. She reaches out with her right

Figure 3-20. A candidate for shoulder and neck therapy.

hand and arm, and because this is fatiguing, she rests the soft forearm on the sharp edge of the pull-out typewriter stand. She also has to twist and lean forward. Her current physical problem is fatigue in the right upper body; a possible future problem may be damage to the ulnar nerve in the forearm.

You can see that the telephone can pose logistical problems—where to place it on a work surface that already holds a keyboard, monitor, mouse, and paperwork. When you assess your own phone from the ergonomic viewpoint, be sure to see how it fits in with other equipment and furniture. It's often the combined effect of several pieces of equipment and their use that produces discomfort, pain, or injury.

Figure 3-22 shows a receptionist. Look at that switchboard.

Figure 3-21. Woman on the telephone making an entry into the computer.

Impressive! Part of the problem is that she has no headset because there wasn't one that was compatible with the switchboard. Upgrading the switchboard would have involved considerable expense, and that wasn't possible at the time. This woman constantly had to twist and lean way forward to get to the switchboard buttons, and so her upper back hurt her all the time. Also notice that her lower legs are quite short—too short for the chair—so they dangle in space. And we know what *that* can lead to.*

She eventually got her switchboard repositioned to the center of the table. She also got a better chair and a footrest, and now, even without the headset, her problem is greatly reduced.

The Keyboard

The construction, placement, and use of the keyboard are very important, so far as maintaining your health is concerned. Al-

*If you've forgotten, turn back to the section on "Chairs and Other Body Parts," which starts on page 29.

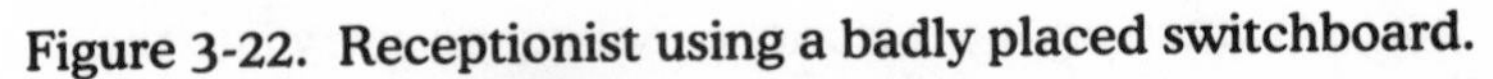

Figure 3-22. Receptionist using a badly placed switchboard.

though firm evidence isn't available, keyboards, and improper keyboard use, probably account for the most severe repetitive strain injuries in the office today. One reason for this, as you might suspect, is just frequency of use. Our fingers are tapping away at those keys more than we think—just estimate the number of words you produce each day in using E-mail, producing reports, doing spreadsheets, or whatever. Then multiply the number of words by five (the average number of letters in an English word), throw in the number of times you use other keys (such as the space bar), and you'll probably be surprised at the result. Word processing personnel can easily exceed fifty thousand keystrokes per day.

It usually takes more than just repetition to cause trouble.

Keyboard strokes by themselves rarely if ever lead to injury. Along with repetitive finger use, we often put our hands and wrists in positions that can be very harmful over the long term. Not only may we not know what a correct position is, the furniture and equipment we use may *force* our hands into dangerous positions.

Incorrect hand positioning can lead to a very serious condition known as *carpal tunnel syndrome*. Many of us have heard of this condition, and some of us may either have had the misfortune to have suffered from it or know someone who has. Few of us, though, know enough about it to take the preventive measures that are absolutely required to avoid it.

So let's take a few moments to look at this condition. What on earth *is* the carpal tunnel, anyway? "Carpal" comes from the Latin word *carpus*, which simply means the wrist. The wrist has a passage in it that really does function as a tunnel. Its walls are made up of bone and ligament, and through it pass nine tendons, used to make your fingers move. The median nerve also passes through the tunnel, next to the tendons. Figure 3-23 shows a simplified view of the carpal tunnel and all the body structures that pass through it.

Figure 3-23. Simplified view of the carpal tunnel.

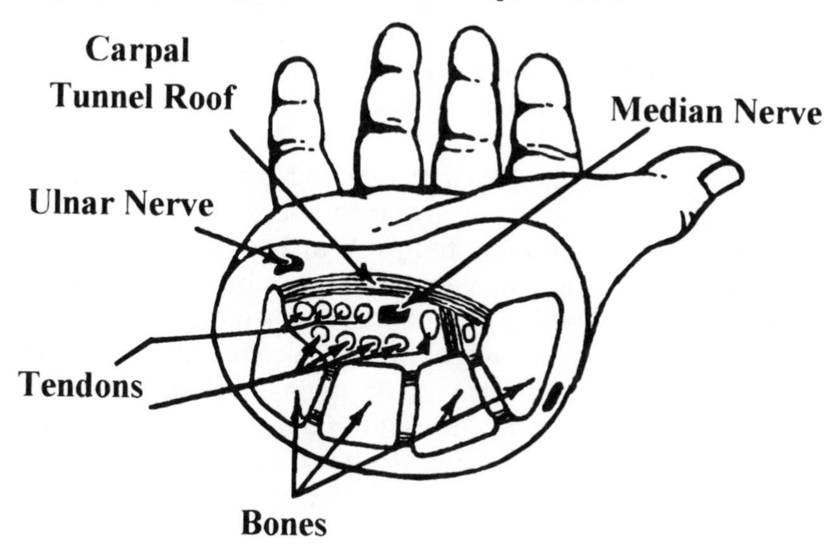

Source: Courtesy of illustrator Paul Schaffer

Because all these things are bunched together, and because they don't have a lot of protection, they are quite vulnerable to damage. If you've ever tried the sport of in-line roller skating, you know the emphasis that is placed on wearing wrist protectors—even ahead of a helmet or knee pads, if you had to choose. That's because of the severe risk to the wrists for anyone falling forward.

The office risk to your wrists, and hence to those tendons and the nerve that pass through the tunnel, can be just as great as that for a novice in-line roller skater trying out a cracked strip of sidewalk. The only difference is, the roller skater gets to the intense suffering instantly. In the workplace, the damage takes a lot longer to build up, and you can be unaware of much of the process until late in the game.

When intense use of the computer is part of the normal work day, a number of factors—wrist angulation, wrist extension, hard-surface pressure—can alter the easy motion of the tendons within the tunnel. Over time, this produces tendinitis, swelling and irritation of the tendons. Fluid accumulates in them, and they swell. These swollen tendons take up more tunnel space and cause the nerve to become compressed; this results in a compromise of the very fragile blood supply to the median nerve, which in turn causes abnormal and distorted messages to and from the hand and fingers. The result is problems with sensation and motor function—and the start of carpal tunnel syndrome.

How bad can it get? Well, before you get pain, you sometimes get a tingling or aching sensation, especially at night. You may find yourself dropping things. Your hands or wrists may swell or feel swollen, and you may lose coordination in your hands. As the pain builds up, you may lose joint movement and find yourself unable not only to work at a keyboard, but even to perform simple functions of everyday living, including driving, opening doors, and cooking. Symptoms don't tend to follow the same progression in every case, but however they appear, they can add up to a progressively devastating loss of your hand function.

This scourge is getting more prevalent all the time. In Chapter 9 we talk about what to do if you think you may be affected by it.

For now, though, we want to show you some of the ways in which carpal tunnel syndrome is caused, so that you can avoid them. Risky hand and wrist postures can be easily identified.

Hand and Wrist Position

Figure 3-24 shows one of the "normal" hand postures of a self-employed architect and successful author who does a lot of typing. The heels of his palms are pressing down on the edge of the keyboard, and thus he's putting pressure—and has been for many years—on the end of the carpal tunnel.

Like many other people who use computers intensively, he also has a pet variation that is equally dangerous. Since he doesn't have a mouse and does lots of editing (requiring moving around on the page), his right hand spends a lot of time visiting the arrow keys. These are, of course, all bunched together. In Figure 3-25, we can clearly see how his thumb is curled under, making his hand into a semifist. Gradually he developed numb-

Figure 3-24. Improper wrist position at the keyboard.

Figure 3-25. Curling the thumb under while using the arrow keys.

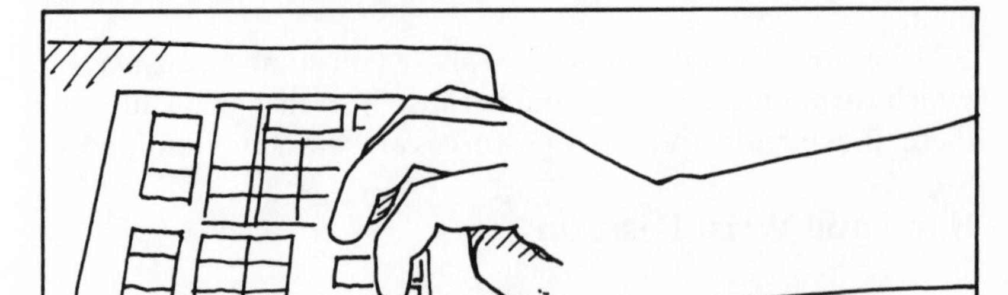

ness and fatigue—but no pain or significant weakness. If you tuck your own thumb under, holding your wrist out straight, and then try moving your fingers quickly, you'll feel the tension in the wrist.

Eventually this man's condition worsened. He has had two operations, but because his was an unusually severe case, today he cannot shake hands without pain, and he must wear a wrist splint most of the time. He is 48, a self-employed professional who is temporarily disabled to the point where he cannot do his job.

We came across a computer user who got into the same sort of wrist position trouble as the architect did, except that here the problem was the disparity between the height of her chair and that of the table: Her chair wouldn't go higher, and the table wouldn't go lower. So her wrists paid the price, making firm contact with the hard, sharp edge of the table as she keyed. It wasn't comfortable, but it wasn't a big deal—until she began to feel wrist pain, her fingers began to tingle, her hands hurt her at night, and she lost hand power. She had carpal tunnel syndrome in both wrists.

She had to keep working. You can see in Figure 3-26 how she attempted to cope, working with protective splints on both wrists. But what's wrong with what she's doing?* She's still getting contact pressure on the wrists, here from the injudicious use of a wrist rest.

We don't want to bore you by harping on the same theme, page after page. But we feel that it's critical that you be able to spot dangerous situations affecting the wrists and hands in your own office, so we'll show you a couple of other, rather depressingly common, situations that result from a mismatch between employee, furniture, organization, and task.

Figure 3-27 shows the arms of a man who went out and acquired an adjustable keyboard holder, but used it the wrong way. He had it at an angle, and he didn't lower it far enough toward his lap. He developed considerable hand and wrist pain. After he saw a photo of himself at work, though, he made the necessary adjustments—and his symptoms improved greatly! By lowering the entire holder, and by dropping the back end of

Figure 3-26. Keyboarder with protective splints on both wrists.

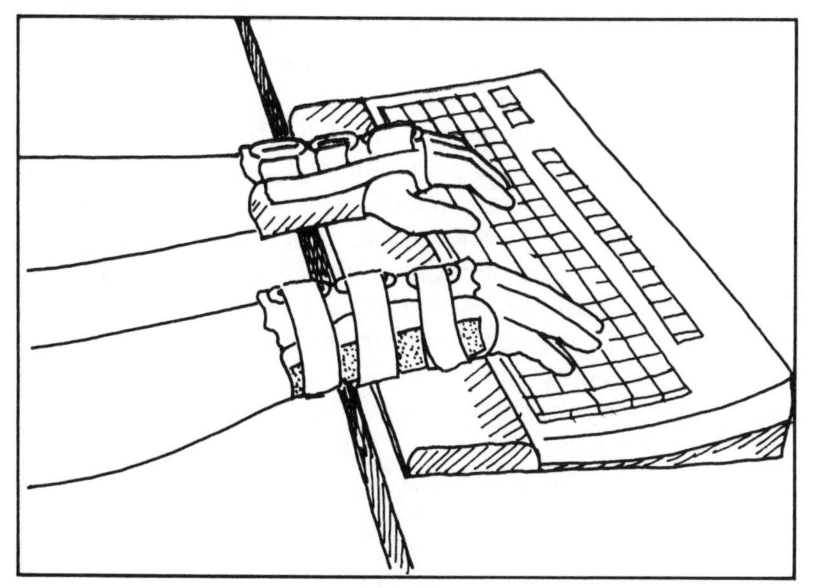

*The table is still too high, and the chair too low. And now she has to stretch her wrists over a wrist rest to use the keys.

Figure 3-27. Using an adjustable keyboard holder incorrectly.

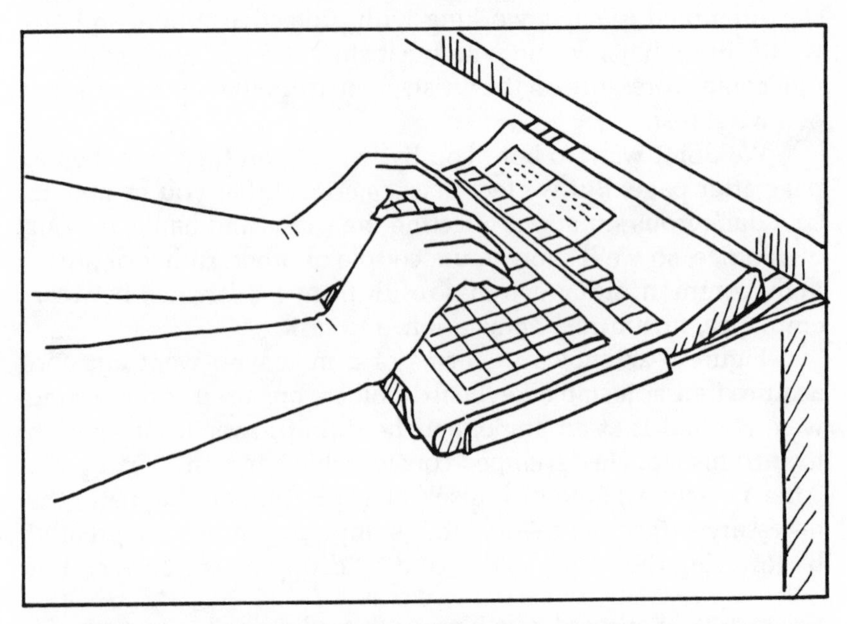

the holder so as to place the keyboard in a neutral position, he was able to work with straight wrists. If you acquire equipment designed to protect you, be sure to read the instructions or get professional help in installing it.

It's not just sharp edges that can cause this sort of injury. A woman in an insurance company (Figure 3-28) had her keyboard too far away; she compensated by keying with her angled wrists resting on the smooth surface of the desk. She developed pain, so she bought a wrist rest. But her wrists were still angled—and they still hurt. Pressure plus angulation always lead to pain. Her friend, who worked in the next cubicle and who had a similar setup, experienced no pain at all. Why not? An analysis of what the friend actually did all day revealed that she had a wide variety of tasks, some of which forced her to get out from in front of the keyboard and move around. So although she was seated a little too high, she didn't get into trouble. And she won't in the future—even if she stops moving around the office—because she's getting a new chair that will let her sit a little lower.

It's very common to have the keyboard placed too far away,

Figure 3-28. Woman with keyboard too far away from her body.

and also to use a wrist rest incorrectly—that is, during typing, rather than for resting.

Figure 3-29 shows proper keyboard posture. Notice that the wrists do not rest on anything. Instead, the worker approaches the keyboard the way a pianist does. Compare this to the way you are working. Note: Even if you have a soft "wrist pad," you should avoid keying from it. If your wrists get tired, get up and do some wrist and hand exercises (see Chapter 6). Wrist pads are for resting, not for keying.

Keyboard design may also affect our health. Traditionally, of course, the keyboard followed the design of the typewriter keyboard, with not much thought given to its ergonomic suitability. The principal difficulty with the traditional design is that we must cock our wrists outwards, so that the fingers can key "straight on." Nowadays there are some innovative designs on the market that permit a more natural wrist position, but there are still millions of the old-fashioned kind in use, and being manufactured and sold.

Figure 3-29. Proper keyboard posture.

The Mouse

The mouse deserves its own section in this chapter, because it is the single most commonly used peripheral device—used, that is, in the sense of your having a physical interaction with it. There are a few things to remember about mice.

First, you should have one that fits your hand. There are mice for left-handed people, and if you're left-handed, you should get one of these rather than trying to make do with a right-handed mouse, which will only twist your wrist in the wrong direction. If you have a hand-neutral mouse, remember that there are software controls that allow you to exchange left and right button functions. If you have a small hand, get a small mouse. Go to a computer store and find one that fits just right. What you want is one that minimizes the travel your fingers have to do to depress the keys.

Second, consider a trackball, which is nothing more than an

inverted mouse. The larger the ball, the better. Again, try one of these to see how you like it. Be attentive to the thickness of the frame; if it's too thick, using it can force you to hold your arm more upward than is desirable. Using a trackball may reduce risk of injury to the soft part of the hand just above the wrist joint; mouse users often use this area as a hand rest, and it can quickly become a source of irritation to the skin on the wrist, and even a source of undesirable wrist pressure.

Third, place your mouse or trackball so that you don't have to reach out to get to it. If you don't have such a place, rearrange your desktop so that you do. Take a look at the man in Figure 3-30; he's got to chase after his mouse, leaning forward from his chairback and stressing his wrist severely. He could solve his problem merely by placing his mousepad in a better spot.

Sometimes the source of a mouse problem is the keyboard, especially when it is on a holder; this may force the user to sit back too far from the work surface. Because most keyboard holders don't have room for a mouse, it has to stay on the desk. The user must repeatedly reach far forward to get to it, with consequent strain on the shoulder.

The best place for the mouse is beside your keyboard and

Figure 3-30. Incorrect placement of a mouse and mousepad.

at the same height. This way you avoid having to reach up and forward. You should be able to glide your mouse easily atop the pad, using mostly the forearm, with your elbow close to your side, and your shoulder rotating easily. Misuse of the mouse actually causes a condition the doctors call "mouser's shoulder." It sounds funny—until you develop it.

And now that we think of it, make an effort to keep your mouse's trackball clean. It's removable, or should be, and you can wipe it off, and also clean the socket it fits into.

Also, use a mousepad—it provides the friction this device needs to function effectively.

If you're using a laptop extensively, and it's got one of those tiny built-in trackballs or pointer devices, consider getting a full-size trackball (or mouse) to use with it while you're in your office. Also consider getting an auxiliary full-size keyboard.

The mouse is a useful (in fact, virtually essential) tool. But if it is not thoughtfully incorporated into the layout of the desktop, the next common RSI may well be mouser's shoulder.

Putting It All Together

Figure 3-31 shows all the elements of an ergonomically well-organized workspace. Check your present workspace and see how it measures up!

Ideal Workstation Checklist

☑ *Monitor:* Height is at or slightly below eyeball level.
☑ *Monitor stand:* Brings monitor to needed height.
☑ *Copy stand:* At or slightly below eyeball level, and close to the screen.
☑ *Task light (lamp):* Illuminates document; does not shine on screen or face.
☑ *Keyboard tray:* Retractable; accommodates a mouse as well as a keyboard. *Some* trays are height- and title-adjustable.
☑ *Mousepad:* Located beside keyboard and at same height.
☑ *Chair back:* Height-adjustable; provides good lumbar support.
☑ *Upper arms:* Neutral position, parallel to torso.

☑ *Forearms:* Nearly parallel to floor.
☑ *Wrists:* Straight (neutral), not resting on any surface when keying.
☑ *Armrests:* Adjustable.
☑ *Footrest:* Supports legs and stabilizes the body. Some are tilt-adjustable.

Figure 3-31. An ergonomically well-organized workstation.

Chapter 4
Repetitive Strain Injuries

Alarm Bells

Although repetitive stress injuries build up over time, there are often some signs that things aren't going well. Some of these signs are in the nature of an "early warning system." Others are final warnings.

The most early common indications that your body is experiencing RSI are pain, tingling, and a numb sensation. We're not trying to scare you, but as you read these next pages, you ought to think about whether you are experiencing any of these symptoms.

Pain from RSI may be felt in the neck, the upper back and shoulders, the forearm and elbow, and the wrist and hand. Tingling is commonly felt in the fingers or hand—either the palm or the back of the hand, or both. It can also be felt in the elbow, and from the elbow through the forearm and into the hand. Frequently, pain and tingling occur together in the forearm or hand.

These symptoms can occur at work or away from work. In fact, occasional pain or fatigue is part of most jobs. But it is the recurring and progressive symptom pattern that is ominous. If you feel repetitive pains only while you're at work, they may become more pronounced as the day or week passes. And if they occur away from the job, they may occur at the end of the day or at night. Often they cannot be tied to any particular activity; they've been reported by people who were driving, cooking, or just sitting. They sometimes occur at night, and can interrupt sleep.

Just what is it that is hurting? First of all, it could be the tendons. These are located on both ends of the muscle; they connect muscle to bone. On one end they act as anchors, and on the other, as pulleys. When the muscles are overused, the tendons can become irritated and inflamed. This is the condition we refer to as *tendinitis.* In the office environment, tendons can also get irritated through contact with a hard surface, such as resting your wrist on the edge of a desk. Ever done that? Don't deny it—we all have! The problem arises when it becomes a habit. Then it gets you into trouble. Those are tendons you're leaning on!

The muscles themselves can hurt. They are the power source for the body; they hold us erect, and are used for every physical action. The most common way to injure your muscles is to hold them in a state of contraction for a prolonged time—e.g., holding your arms out over a keyboard without taking a rest. A sort of metabolic "sludge" then accumulates in the muscle tissue, leading to both pain and weakness. Too much sludge for too long can inflame the muscles, leaving them sore to the touch and painful to use.

When you stretch, reverse the prolonged positions, and relax, decontracting the muscle, this opens up the local blood circulation and allows the clearing out of the sludge. You feel good again!

And the nice thing is that you don't have to do these "reverse stretches" for very long. You might have held a particular position with your arms for an hour. But just a short period of stretching is enough to counter the effects of the contraction—a minute or so, if it's done at regular intervals.

As you know, the body is loaded with nerves, which carry the electrical supply we need in order to function; these, too, are pain-sensitive tissues. With prolonged muscle contraction, the nerves can get pinched—a good reason for periodically stopping to stretch the muscles. The second way nerves get injured is through pressure, just like the tendons. This is particularly true of the "funny bone"—the ulnar nerve in the elbow—and the median nerve in the wrist. A more subtle form of nerve injury can appear in the armpit, in the area known as the thoracic outlet. We'll be talking about this in greater depth in a little while.

And finally, the bones themselves have an outer covering, the periosteum, that is sensitive. Again the hand and wrist are often affected. Also, the many interlocking bones of the spine, and particularly in the neck, are vulnerable.

As for the tingling or numb feeling of RSI, it can result from pressure on any of the nerves, ranging from deep nerves, the ones exiting through the spine, to the more superficial nerves found near the surface of the forearm and along the sides of the fingers. And these tingling symptoms are often accompanied by pain if the nerve is injured. Inflamed tendons in the elbows also can produce tingling that shoots to the hands.

There is a common denominator for all the sensations we've just described, as they move from the stage of fatigue to that of prolonged symptoms (that don't disappear when you cease, or change, behavior). This is the development of *adhesions,* which are really scar tissue that forms beneath the skin. With prolonged tissue inflammation comes the formation of scar tissue. This tissue is tough and hard. It can bind over a muscle and alter the tendon structure, so that motion becomes painful. If it should bind over a nerve, it can cut the blood supply; the result can be pain that is truly terrible. Carpal tunnel syndrome is an excellent example of this: The median nerve in the wrist can become bound over by scar tissue.

If you are lucky, the scar tissue can be light, and its effects can be ameliorated by such techniques as deep muscle massage or ultrasound treatments by a physical therapist. If, on the other hand, the scar tissue is thick and uncompromising, then there's no alternative but to entrust yourself to the sharp blade of a good surgeon. It's worth taking some preventive measures, right?

Where and Why It Hurts

The Neck

The neck area is full of muscles and tendons that are often affected by RSI. When pain originates from these muscles or tendons, it is often felt as a spasm—an intense but brief soreness. Easy, slow stretching can often relieve this. But if it does not, and if the spasm persists, it can become a prolonged condition. Now it can bother you both on the job and at home, and it often ultimately sends you to your doctor or chiropractor.

What are the actions that can cause injury to the neck? As you might suspect, any prolonged posture, or repeated action, that isn't "natural." For example, craning your head to look at an offset monitor. Cradling the phone receiver between your neck and your shoulder. Bending your neck so that you can see the keyboard, or raising it so that you can read what's on the monitor. Or leaning forward in order to see what's on the monitor more clearly—perhaps because of poor eyesight, or glare, or just too small a font. All these are very commonly done in the office environment. How many apply to you?

These twisted and angled neck postures also twist the deep neck bones, causing bone-on-bone pressure and pinched deep nerves. Chiropractors see many patients with this kind of pain. This condition is no fun; it makes driving, reading, and even sleeping difficult.

The Upper Back and Shoulders

The upper back and shoulders have the muscles and tendons that support the upper torso and head. This is fairly important work; after all, it keeps our faces from falling forward into our keyboards! Yet this area is often subjected to punishing strain that if allowed to go too far can strongly affect our quality of life. The principal reason for the strain is that, no matter how well balanced we are in our posture, and how well organized we've made our desktop environment, when we work with computers, we inevitably tend to lean forward. It's virtually unavoidable, to some extent. All our work is in front of us—our documents, telephone, monitor, and keyboard. We never find these things behind us! The longer we perform our keyboard tasks, the longer those upper back muscles remain in contraction, building up a load of metabolic sludge. So it's *imperative* to take microbreaks, and reverse those postures by stretching the muscles—

before they start ringing the alarm bells. We'll get into that more in Chapter 6.

In many offices, new technology sits atop old furniture. This frequently results in the keyboard being located on the desk or table surface, too high for the person who has to use it; it may be only several inches too high, and it appears perfectly usable, but guess what develops over time—sore shoulders and necks! Getting the arms up just that inch or two requires relatively little force from those shoulder muscles, elevating the arms just enough to strike the keys rapidly—only a minor inconvenience, so it seems. But having to hold the arms and shoulders up for long periods of time means continuous muscle contraction, metabolic sludge build-up, tightness and heaviness, and gradually increasing discomfort. If you shrug your shoulders a lot during the day, stretching them for relief, you should determine if the keyboard is located too high. A good massage might be in order, until you figure out how to get the keyboard lowered.

Large-bodied people are at particular risk for shoulder injury in today's office. The risk factor has to do with the width of the alphabet key portion of the keyboard, the part you use for letters, names, documents, etc., as compared to upper body size, specifically the shoulder-to-shoulder width. The greater this width measurement, the wider apart are the arms and forearms when they assume a keying posture—upper arms resting along the side of the torso, forearms nearly horizontal, wrists neutral, fingers ready to work the keys.

Thin people with narrower frames have no problem, as the letter-key width is practically the same as their shoulder-to-shoulder width; their forearms can easily come straight forward to the keys. But as a person's frame size widens, the forearms end up farther and farther apart, right arm over the number pad rather than the alphabet keys. Even for those people with "usual body size," this width relationship is a bit off. If you are truly wide-bodied, check out how your arms are located in relationship to the letter keys—we bet you have to scrunch them in toward the middle, and you probably also have to angle your wrists outward (toward the little finger) so you can line up with the key arrangement. There are three possible ways this width mismatch can cause shoulder problems.

The first is when you hunch your shoulders upwards and extend your elbows outward to each side, like wings. The muscles of the shoulder girdle are very powerful, but holding this position for very long is extremely tiring—just try it for a while, without a keyboard or other surface to rest your wrists on. Because of the resulting forearm posture, this additionally requires ulnar deviation of the wrists—cocking each wrist to the outside, to get the fingers lined up properly on the keyboard. This results in sore shoulders *and* sore wrists.

The second way shoulders are affected occurs when you sit back a bit from the keyboard and reach forward, gradually bringing your arms together. Your arms now have to be held in the air; this can be done only for so long before your shoulders fatigue. At this point, you usually relieve the strain by supporting the arm weight on your wrists, on the top or edge of the desk. The result is both sore shoulders and sore wrists.

The third way is squeezing your shoulders into your body, in an effort to make yourself narrower than nature intended. This places strain on the shoulders, and also puts tremendous pressure on the upper armpit. Exiting directly from the armpit is the major nerve and vascular supply for the arm, wrist, and hand. Pressure on this nerve bundle can have painful, and often bizarre, consequences. Depending on which part of the nerve bundle gets compressed, there can be such varied and unpleasant symptoms as pain and tingling in many parts of the arm, radiating discomforts that seem to change from day to day. Also, there can be skin mottling and alterations in limb temperature. One can have the sensation of being cold, even in a warm room.

These cases of unusual arm pain are very common, and very bothersome. They are also very slow to respond to treatment. The condition is often referred to as *thoracic outlet syndrome.* When the cause is removed, the condition usually gets better eventually. No one should agree to have surgery for it unless the appropriate diagnostic nerve tests have been done and a specialist has been consulted.

One solution is a special keyboard. Some newer keyboards, with both sides canted outward from the middle, may prove helpful in addressing this situation, since the wrists don't have to be angled or the armpits compressed.

Many office workers, regardless of body size, must continue to function with the keyboards they now have. Is this reason to despair? Not at all. There are exercises you can, and must, include as part of your exercise routine if you use a standard keyboard. Chapter 6 gives you all the information you'll need to do this.

One of the more common causes of *shoulder* strains and pains is reaching forward for the mouse, holding it and manipulating it with the shoulder muscles working all the while to keep the arm up and forward. Medical and ergonomic literature is beginning to describe a problem called "mouser's shoulder," as mentioned in Chapter 3. The cause of the problem has to do with available space and with awareness on the part of the employee. Spatially, there is the tendency to place a mouse beyond the keyboard, or well away from the keyboard; both setups necessitate a reach. When undersurface keyboard trays and holders are used, there is no room for the mouse, so it ends up on the desk or table, or on a side shelf, and again reaching is required. The solutions are to organize better when space permits and to request a keyboard holder or attachment that will provide support for the mouse.

Forearm and Elbow

There are muscle and tendon structures on the back and front of the forearm and elbow. On the back of the elbow, the anchor tendons (extensors) all hook onto the same area of bone; that is, they cluster on the outer part of the elbow (lateral epicondyle). In sports, irritation of this area is referred to as "tennis elbow." Similarly, on the front of the elbow, the anchor tendons (flexors) are likewise clustered, but on the inner bone of the elbow (medial epicondyle). In sports, irritation of this area is sometimes called "golfer's elbow."

Of course, medical doctors aren't content to use sports terms, so these conditions are called by your doctor, respectively, lateral epicondylitis and medial epicondylitis. (You will not be quizzed on these terms.)

An effect of tennis elbow (lateral epicondylitis) may be pain when you do such a simple thing as taking a jug of milk or juice from the refrigerator—you use your extensor muscles to do this. Even holding a cup of tea or turning a doorknob can become a chore—or impossible—in an advanced case. And in very severe cases, turning from side to side in bed and putting pressure on this area can result in intense pain. Hitting it against a door jamb isn't even to be contemplated—it's too painful to think about, like biting your tongue repeatedly, though it does happen to those who suffer from the condition.

There's a lot of tennis elbow coming out of offices these days. How does it occur? Generally speaking, this condition develops over time when the extensor muscles are overworked, having to hold up the wrist and at the same time raise the fingers after striking the keys. The more the arms reach forward and upward, the more pronounced this becomes, so a mismatch between chair height and work surface height, requiring a person to reach and then hold the arms up to the keyboard, results in metabolic wastes accumulating in the extensor muscles. As wastes build up, the muscles hurt and function less well, so their tasks are shifted to the anchor tendons in the elbow, and the process of tendinitis (tennis elbow) begins.

Another cause of tennis elbow is placing the keyboard at a distance from the body; in customer service jobs, for instance, employees may place statements or other documents flat on the surface, between their torso and the keyboard, so that they can write down numbers or identify data on the paper as they proceed in their task. Thus, to use the keys, they extend the forearms and reach to the keys, using those extensor muscles to hold or cock up the wrists. This is more pronounced as they reach back toward the further keys and the function keys, and it is more pronounced if the back legs of the keyboard are in use, angling upward the back of the keyboard and requiring an upward reach as well.

The combination of writing and keyboard work often presents an ergonomic predicament, as there is only so much space available for work arrangement, and neither the keyboard nor the paper can be easily eliminated; separated sloped writing surfaces may help, as may undersurface keyboard holders, shifting the task elements in different three-dimensional configurations. Long-term solutions are encouraged, but we recognize that they are not simple and may require significant investment of time and money; solutions generally involve restructuring job tasks and using technology more creatively as it evolves.

There isn't as much "golfer's elbow" coming from the office environment, because there are few activities that place strain on this area. However, using thumb and finger force on the keyboard, involving power use of the flexor muscles, can lead to this type of elbow pain. Constantly and tightly squeezing a pen or pencil, or a mouse, also involves the flexors, and can likewise get you into difficulties. Finally, these flexor tendons anchor the flexor muscles of the forearm, and the tendons at the other end pass through the carpal tunnel. Irritation at the wrist end can actually cause secondary irritation at the other end (the elbow).

It's not terribly difficult to do a self-diagnosis to see if you have these conditions—if you know just where to poke, that is. Don't hurt yourself by doing this too hard, but if you hold your arm with the palm up, and feel firmly with your thumb in the inner "hinged" part of the elbow, and if you feel pain, chances are you've got golfer's elbow. For tennis elbow, hold your arm out with the palm down, and feel at the top of the joint. These conditions are fairly common, since many home activities can also cause them. When they get past a certain point, however, you'll need medical advice. You'll also need to change some work (or play) habits.

Moving beyond the elbow, we come to the forearm—a part of the upper body that is often subject to RSI. We tend to rest the soft tissues of the forearm on a hard, or even sharp, surface—the edge of a desk, for example. And we do this not just at the office, but in restaurants, while standing at counters, and in many other places. This puts pressure on the flexor muscles and can result in muscle pain. Of course, in the office these forearms aren't just resting, they are working the keyboard, and that means the muscles are contracting and relaxing, and the tendons in the wrist area are moving rapidly back and forth as the fingers go up and down. There is a lot of motion occurring, and when the forearms are on a hard edge, the pressure of the edge produces resistance and friction forces against which these moving muscles must work; they have to work harder, they fatigue more quickly, and eventually pain develops in these flexor muscles as metabolic wastes accumulate, the muscles become swollen, and tissue fibrosis begins to develop.

If the desk edge hits against the far part of the forearm, closer to the wrist, then tendons get involved as well; the flexor tendons extend back some 3 or 4 inches from the wrist into the forearm, and they are close to the surface. Desk edge pressure here can produce tendinitis, and can endanger the carpal tunnel region of the wrist. This tendinitis injury is usually much less forgiving than simple muscle pain, and it can produce a painful condition involving the entire underside of the forearm, eventually extending the process into the flexor muscles and all the way back to the elbow (medial epicondylitis).

Take a look at your work space and what you are doing with your forearms. If you are subjecting them to those sorts of stresses, change something, anything—but stop doing it. The price is too high.

The Wrist

Everyone has heard of carpal tunnel syndrome (we've mentioned it in several places in this book already). This is where it happens—at the base or the heel of the hand (Figure 3-23), beneath the wrist crease. Within this area there's a very tiny space, which is crowded with nine tendons and one important nerve (the median nerve). This nerve becomes pinched by inflamed and swollen tendons (flexor tendinitis), scar tissue, or both. The effects of median nerve injury include pain, tingling, and numbness in the wrist, in the palm of the hand, and in the thumb and the index and middle fingers. If the process persists and progresses, it can result in loss of muscle at the base of the thumb, and resulting weakness. When this muscle atrophy occurs, the solution is often surgery.

The majority of cases of wrist tendinitis do not result in carpal tunnel syndrome. Most of them do result in pain or tingling in the wrist or the palm of the hand. Regardless of whether the damage is to the tendons or the nerves, however, and regardless of the severity, the office mechanisms of injury are always similar, and can be identified.

The most common situation that causes wrist tendinitis or

carpal tunnel syndrome is that while the employee is keying, the wrist rests on the table or desk, and the hand is cocked upward (see Figure 4-1). This can cause such intense pressure on the tunnel that it chokes off the fragile blood supply to the nerve by literally collapsing the blood vessels. This can start the process leading to carpal tunnel syndrome. Even without involvement of the nerve, however, such a wrist posture puts pressure on the tendons, and can cause painful flexor tendinitis.

The second most common cause of flexor tendinitis is wrist angulation (Figure 4-2)—holding the wrist at an angle, either outward to one side or up or down. While the tendons move freely when the wrist is in a neutral, "straight ahead" position, as soon as it is cocked in any direction, the tendons encounter the resistance of friction as they pass over the bones of the wrist. Straining repeatedly over the wrist bones, the tendons gradually become swollen and inflamed.

Most people believe that it is the repetitive movement of the fingers that causes carpal tunnel syndrome. We doubt that finger movement alone is the culprit. Instead, it is the finger movement

Figure 4-1. Resting the wrist on the edge of the desk while keying.

Figure 4-2. Wrist angulation.

coupled with improper wrist alignment, often with the added factor of contact with a hard edge. This isn't to say that repetitive hand and wrist work doesn't pose any problems, even if there is proper alignment. Pure speed in working on the keyboard, without "pace and grace"—without taking breaks, stretching the hands and fingers, and using just the right amount of force—does not make for a more successful keyboard operator. In fact, ergonomic studies consistently show that after several hours of uninterrupted keyboard work, the error rate increases significantly (this is probably the result of the accumulation of metabolic brain sludge in addition to physical fatigue).

Also, as Chapter 2 points out, our bodies, including our wrists, tend to accumulate injuries from the normal events of life. The wrist may have scar tissue, which can make the tunnel tighter to begin with. Hence the person is more vulnerable to carpal tunnel syndrome—there is less room in the tunnel to accommodate the milder cases of tendon swelling, and so the nerve is more easily compressed. In Chapter 6, we talk about how to do periodic "flight checks" on your work posture, including the wrists. That, plus "ergobreaks," should form a normal part of every work day.

Moving now to the back of the wrist (yes, that's vulnerable, too!), we come to a group of tendons known as extensor tendons. These are what give us the ability to raise our hands and our fingers. They are vulnerable to extensor tendinitis, which results from angling the wrist upwards (dorsiflexion) and performing repetitive finger motions. Laying the forearms along a desk surface and cocking the wrists upward to key is a common way to develop this condition.

Finally, the joints of both the front and the back of the wrist and of the fingers are fairly frequently prey to ganglion cysts—little lumps and bumps. They're something like the little swellings that can occur on a car tire when it develops a weakness in the sidewall. The joint "blows a bubble." These may come and go, and reappear; they may move around, and sometimes they grow in size. If they get bigger, they can cause pain that increases with wrist and finger motion. They're filled with joint and tendon lubricating fluid, and so they're squishy. The standard medical treatment used to consist of slamming them with a hard surface, such as a book, to disrupt and destroy them. (Medical history does not record which titles were the most effective, but the Bible was commonly used—really!)

If these become large and painful, it is time to see an orthopedic or hand surgeon. Sometimes the fluid can be drained using a syringe and needle; this may work but often the darn fluid reaccumulates with time. Surgical removal of the ganglion is the next option, depending on the amount of pain, interference with function, and cosmetic considerations of the hand and wrist.

The Hand

Pure hand pain, as distinguished from pain that radiates into the hand from the forearm or wrist, usually comes from stress and strain put on the fingers and the thumb.

Thumb

You may wonder how on earth one can misuse the thumb while using a computer. Trust us—it's often the case with certain number pads. Take a look at Figures 4-3 and 4-4. In Figure 4-3, we can see the worker using the number pad. She's tucking her thumb under the rest of the hand so that she can use it to strike the "0" key. This works on many calculators where the "0" key extends leftward beyond the margin of the "1," "4," and "7" keys, and doesn't even require much of a tuck at all, but on most standard keyboards, that "0" key is directly lined up along the left margin of that number pad, and habitual use of the thumb for an "0" does require the tucking. This translates into continuous use of the thumb muscles, producing thumb pain and also increasing pressure within the carpal tunnel space. A safer approach to the number pad would involve using the index finger for the "0," leaving the thumb out of the action.

In Figure 4-4, the worker is again using the keyboard, but this time the thumb is splayed so as to reach a special key—

Figure 4-3. Tucking the thumb under when using the number pad.

Figure 4-4. Splaying the thumb in order to reach a special key.

either the "enter" key or the space bar. This position is called "splay thumb," or sometimes "space bar thumb."

Both of these are common behaviors, and upraised thumbs are just as common. Again, it's as if the thumb were an embarrassment to the rest of the hand. We want to get it out of the way, and so we lift it up. And this leads to tendon and muscle fatigue.

Another cause is pounding the space bar with the thumb. That causes pain at the base of the thumb from the forceful use of the thumb muscles.

The activities mentioned above are, in the aggregate, frequent enough that thumb tendinitis has become extremely common. It is quite painful, and can take a long time to clear up.

Fingers

Let's start with an appeal, or perhaps a warning, to those who take pride in having extra-long fingernails—real or artificial. Sure, you've been able to handle your keyboard just fine, much to the amazement of your colleagues. But those talons have forced you to work the keys with your fingers straight, rather than flexed. You've been unconsciously increasing the amount of work the flexors and extensors of your forearm have to do, just to protect your nails.

You can actually sense this for yourself. Hold your arm out straight, and move your fingers rapidly up and down in the correct keying position—that is, working with the tips, with the fingers flexed. There's a sensation in the forearm as the fingers move up and down. Now try it with your fingers out flat. Feel the increased pressure and forceful movement in the forearm? You're using both flexors and extensors at the same time—a lot of work for the forearm, and a good way to get a repetitive strain injury.

Just as there are thumb bangers on the space bar, there are also finger pounders. This is really jarring to the joint structures, the finger tendons, and the sensitive nerves that run along the fingers. It's usually not necessary to use so much force. If it is, then you should change your keyboard for a better one. It is more than likely that forceful keying is a bad habit that needs to be broken.

Some workers are double-jointed in the fingers. That means that when a finger strikes a key, it may bend outward. To compensate, such workers often make a little extra effort: a slight movement of the wrist, both up and down, creating a much more forceful key strike. Many thousands of repetitions of this movement can cause hand, wrist, and forearm pain. What can you do about this? If your keyboard has keys that are pressure-adjustable, you might try making the touch lighter. Or you may need a different type of keyboard. In any case, if you are double-jointed, be aware that you may be using too much force, even though it doesn't seem that way.

The little finger comes in for its share of abuse. Especially while working on the number pad, some workers tend to extend this finger while flexing the others. This can cause pain in the forearm—and, believe it or not, in the elbow. Remember the condition we talked about above called lateral epicondylitis? And why the elbow? Because that's where the extensor tendons of the little finger are anchored. Therapists called this condition "pop-up pinkie." It is similar to the position in which some people put their little fingers when holding a glass of wine.

These are such tough habits to break that sometimes therapists will tape fingers together to prevent them from "misbehaving."

Other Hand Problems

As we have seen, placement of the mouse and mousepad is dependent on space and may not receive much forethought. Mouse function is based on the two-dimensional orientation of the mouse and the direction it is moved. If the mouse is too far away from the operator, the wrist can easily become kinked at the base of the thumb while guiding the little critter; on a recurrent basis, this leads to hand pain where the kink occurs.

Also, double-clicking can take its toll on the index finger and the back of the hand and wrist. As we mentioned earlier, there are a great number of mouse designs, some with badly positioned clickers that require awkward finger postures and lead quickly to hand and finger fatigue. Clicker sensitivity is controlled with software commands, usually in the control panel of the main menu; not all mouse users are aware of this option. Finally, many mouse designs incorporate at least three clicker buttons, with a software option that assigns a double-click function to a single-click stroke on a particular clicker button. That's right, never double-click again!

Another device we mentioned, the trackball, has its own design considerations that can affect the hand and fingers. For example, the size and location of the ball vary considerably, and some designs appear to result in thumb pain from repeated manipulation of a left-located ball. Also, thumb pain has been seen from repeated use of a left-located clicker button, straining the thumb abductor tendons.

Finally, hand pain involving the thumb and index finger can arise from a lot of writing, especially if duplicates or triplicates are used, and one has to bear down hard. Technique has a lot to do with "writer's cramp"—specifically, how tightly one holds the pen or pencil; lefties have the additional tendency to angulate their wrist as they write. Many times this can be avoided by using a large utensil—the fatter the utensil, the less forceful the pinch grip used for writing. There are a number of styles available of slide-on adapters, which effectively expand the diameter of the utensil, again reducing the pinch force used. Some people find that adding a sloped writing surface (available in a variety of styles) makes writing tasks easier.

Putting It All Together

We have isolated anatomically the various components of the upper body and extremities, in order to define what hurts, and where. We wanted to show which body parts are susceptible to injury, and which ones get stressed in the office environment. In fact, pure injuries are the exception rather than the rule. For example, if a tendon hurts, the muscle attached to that tendon will probably also hurt. Then the forearm and wrist may be painful when examined.

When a nerve is injured, it can create "upstream" and "downstream" sensations, with pain shooting all over the place.

And when the shoulder hurts, the elbows and wrists can be secondarily stressed.

Repetitive strain injuries are generally not simple. But they can be diagnosed, and if caught in time, usually they can be cured.

Chapter 5
The Eyes

Eye complaints are frequently number one on the list of workplace problems mentioned by employees. We'll tackle these by looking first at the piece of equipment employees spend a lot of time looking at—the monitor.

The Monitor

Monitors may present some special vision problems in the workplace. One important step to take to avoid these is to be sure that you have a good monitor. What defines a good monitor? To understand the question, let's first take a look at Figure 5-1, which shows a cross section of a typical CRT (cathode-ray tube) monitor. This is a side view of the guts of that thing you stare at all day.

Notice the little device called an electron gun. What this gizmo does is actually shoot a stream of electrons out toward the screen. When these little guys make contact, they hit a phosphor coating on the inside of the screen, and cause a little bit of it, called a *pixel,** to glow. "Pixel format" is a phrase that simply means the way these little elements are laid out on the screen in rows and columns. Older PC pixel formats tended to be limited—say, 640 by 480 pixels. With newer software, that's very hard on the eyes. A better PC format is 1,024 by 768; it's much sharper. High-performance workstations can have pixel formats of 1,280 by 1,024, or even higher. Obviously, these are easier to work with, and, naturally, they are more expensive.

*Short for "picture element."

Figure 5-1. Cross section of a CRT.

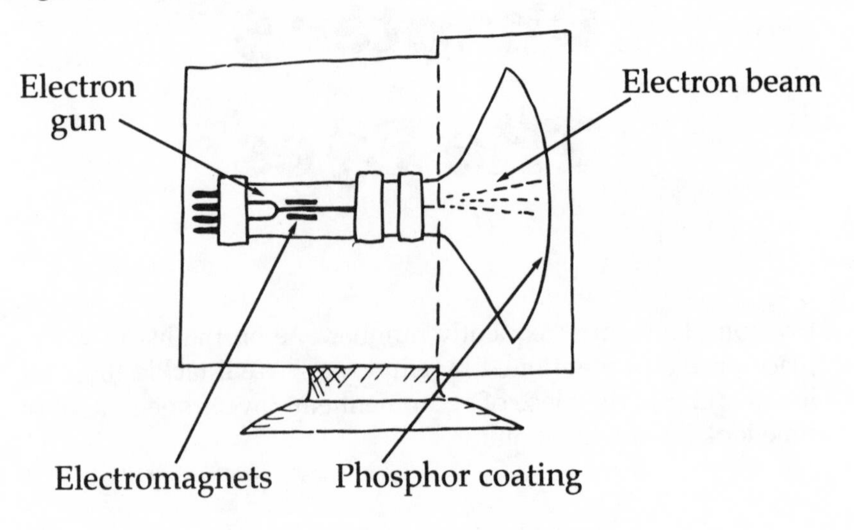

For the sake of simplicity, our sketch shows a "grayscale," or black-and-white, monitor assembly. If this were a color monitor, there would be three electron guns, one for each of the three primary colors, each one shooting out its own stream of electrons. The coating on the inside of a color monitor screen consists of *triads* made up of red, green, and blue dots.

Now the target for those little electrons just became that much harder to hit. Those little dots also have to be placed just so if the focus and color are to be clear, not smudgy. If they are not clear, you've got a case of *misconvergence.* If you have misconvergence—fuzzy or smudgy color—change your monitor. The same goes for finding color fringes around the text. This isn't caused by deterioration of the monitor as a result of its age; it's usually a question of the quality of the monitor and, notably, of the accuracy with which the *shadow mask* (a screen with holes in it that focuses the electrons onto the screen) is manufactured and installed.

Moving right along—and this is where we get back to *you*—

the stream of electrons doesn't spray the whole screen at once. Instead, it moves over the screen, rather like those lawn sprinklers that sweep jets of water to and fro, again and again. The rate at which the electrons sweep across the screen time after time is called the *refresh rate,* because the electrons are refreshing the glow of the little pixels. The faster the refresh rate, the clearer the picture or words on the screen will be. The refresh rate is generally expressed in *hertz,* which isn't an auto rental company in this case. It is a unit of measure named for Heinrich Rudolph Hertz, who discovered that light comes to us in waves. So, for example, a 76-Hz (for hertz) refresh rate just means that the electrons are sweeping across the screen 76 times per second. And that's excellent; lower is worse, in terms of one of the important elements that affect your eyes. The standard used to be 60 Hz. At that refresh rate, you can detect flicker. Letters and images jump just a little bit, but at a very rapid rate. This isn't desirable at all.

So one way of finding out how good your monitor is, is to find out the refresh rate. How do you do this? Ask the person in charge of your computers—he or she knows, or can find out.

A good monitor is also one that has high luminance and high contrast, features that result in a crisp, legible display. *Luminance* is what you get when the electrons strike the coating on the inside of the screen, and those little pixels light up. The problem is that like some of us, the coatings tend to lose their zip as the monitors get older. When this happens, your eyes have to work harder than they should to read what's on the screen. Not good.

So how do you tell whether your monitor's screen coatings are wearing out? Well, there are ways, but we'll bet the office doughnut budget that if we were to tell you how to do it, you never would. It's too complicated. So, if your monitor has been hanging around the office for five years or more, it is very likely to have lost significant luminance, and you should have it replaced.

Contrast is, in part, a function of the accuracy with which the electrons are focused onto the screen, as we've already mentioned. But the *pitch* of the monitor also comes into play here. Pitch is the distance between individual dots, or pixels. To determine pitch on a color monitor, where the dots are in triads, the measurement is simply made between two dots of the same color. The smaller the distance, the sharper the image. So a monitor with a pitch of 0.28 (of an inch) is sharper than one with a pitch of 0.33.

The front of the monitor screen should be as flat as possible, so that lines appear straight, not curved. If you notice curvature other than at the very edges of the screen, get another monitor. And the screen itself should have an antiglare treatment, to reduce interference from other light sources in the work environment. Newer monitors generally have this treatment, but you can check by asking your computer specialist for a look at the monitor specs, or product specifications. If no one knows where they are, write down the make and model number from the back of the monitor, find out where the manufacturer (or the nearest sales office) is located, and call and ask. Many general computer stores will also have access to this information, if you can get the attention of a sales clerk.

In what we've written above, we've advised you to change your monitor if certain conditions exist. You may think that this is going to be difficult to sell to your supervisor, to the facilities people, or to whomever. The natural tendency is to let things just slide for another five weeks. But don't. Unless you're only temporarily visiting us from another planet, you have, and will have, only one pair of eyes to last you all your life. It's worth a little effort, if you use a computer intensively, to get the best monitor you can.

Monitor Placement

Monitor placement with respect to *your* placement, as well as to light sources, can also affect your eyesight. So let's make a quick check of how your workplace shapes up. Many of you spend a lot of time staring at the monitor screen. When you do this, it's just like spending many hours reading books, so far as your eye muscles are concerned. Just as you focus on the letters in a book, your eyes have to focus on what's on your computer screen. Eye experts think that when you do this a lot, your eyeball actually changes its shape over time. It does this to get the best focus for the work you're making it do. This kind of remolding of the eyeball produces myopia, or "nearsightedness." Not a fatal disease, to be sure, but it does throw off your vision at other distances. Of course, a degree of myopia can be expected to come along with aging. But if you're in your twenties, why hurry the process along?

One way to lessen the impact of prolonged staring at the screen is to enlarge the characters, either by a program command or by using glasses or a magnifier in front of the screen. Some optometrists recommend glasses that provide low-level magnification for all people who work with computers.

Another way is to take a break every hour—take a stroll outside and look at the trees, or just look out a window. Our eyeballs have a rather nifty system of lubrication. Many of us know what it's like to have dry eyes—it's not comfortable, and if we're wearing contact lenses, it can be painful and harmful as well. Today's office tends to have dry air, and staring for prolonged periods at a computer screen also produces dry eyes. This condition can be considerably worse for people who take certain medications, the most common of which are over-the-counter antihistamines.

Another potential source of problems is the addition of color to the monitor screen. Color vision, or acuity, plays a key role in ease of viewing. You'll want to make color changes to the default setting of your monitor if you have any color vision problems. Technical support people should be able to assist you.

Working with a computer requires attention both to the monitor screen and to hard-copy documents on your desk or in some sort of document holder. Your eyes are almost as busy as your fingers. Tasks carried out on the computer, though easier for us in some ways, are arguably as eye-intensive as the reading and hand copying of documents a century ago.

The sense of vision feeds the brain details, which the brain must then process for correctness, style, format, and so forth. Visual fatigue can lead to brain fatigue, and thus to all sorts of errors in our work. It doesn't make the workday any shorter, either. So from the start, it's wise to have insight into our eyesight.

Since viewing the monitor screen is like reading, our eyes

focus for long periods of time at the same distance. Eyes operate, of course, with the aid of muscles; these pull on the eyeball both to change direction and to change the shape of the eyeball. This change in shape is what enables our eyes to focus at different distances. Eye muscles can fatigue with prolonged use, just like any other muscle; this produces metabolic sludge, which can result in a significant headache around the eyes and eyebrows.

Now, characters and colors on a monitor screen are not fixed images, but dynamic images. As we've seen, they are constantly refreshed—that is, redrawn. Older systems have slower refresh rates, which, as we've seen earlier, can produce a noticeable flickering of the image. This is not a healthy situation. If that's what you're getting on your monitor, change your system. Generally speaking, older people are more sensitive to flicker caused by low refresh rates.

Many systems allow you to adjust color combinations and font size, which can increase the clarity of the image on your screen.

Office Lighting

Many modern offices have a combination of internal and external light sources. Internal light can be general, that is, lighting that illuminates the whole office; general lighting is almost always overhead, and is usually fluorescent. Or internal light can be "task" lighting—lighting that is usually on a counter, or from a lamp designed to throw light onto a document.

Both types of light sources can result in glare and reflections from the screen and from the surface of printed materials. This is something to guard against, because the pupils of our eyes have a reflex tendency to constrict. Constriction alters the focus of the eyes and disturbs the visual process of reading characters or graphics; the pupil needs to dilate in order to read the screen. The presence of glare, therefore, puts the eye under considerable strain, making it want to constrict and dilate simultaneously.

Your monitor screen can also display glare through excessive brightness or inadequate contrast settings. Another source, less prevalent in newer workspaces, is from paneling. In better-designed environments, cubicles and office walls are in some nonglare, neutral color.

In addition, it's surprising how often office lighting is found to shine directly into the eyes of office workers. Some workers try home-grown methods of shielding themselves, from wearing caps with visors to constructing elaborate cardboard shading devices, or even to erecting an open umbrella.

In areas where there are outside windows and some work positions face toward the window, there may be clashes between employees who want the direct light and those who suffer from it. This light can reflect off your screen and wipe out part of the image; this can induce you to twist and crane your neck in order to see better. When taking remedial measures creates conflicts with your coworkers, remember: The eyes have to be protected, and if that requires moving to another part of the workspace or building, so be it.

Protective Strategies for the Eyes

As a rule of thumb, vision testing should be performed every two to three years. If it's been longer than this since you've had it done, get going—make that appointment. And when you do see your eye doctor, be sure to report that you work with a computer. Doctors sometimes prescribe low-level magnifying glasses, as we've already noted, for all persons working with computers. Sometimes a mild tint—pink or gray—or an antiglare coating is also prescribed.

Ask specifically about your ability to discriminate colors and shades of color. Know about any eye weaknesses or wanderings, as this can influence how quickly your eyes fatigue. Inquire about what, if any, special measures you should take if you wear contacts, strong lenses, bifocals, or trifocals; this can affect monitor placement and your need for a higher refresh rate or a particular dot pitch.

Tell your doctor if you've noticed any problems with dry eyes, headaches, or eye fatigue. If you notice that you're having trouble reading road signs while driving at night, report it; this can be a sign of early myopia. (Wearing a pair of low-level magnifying glasses while using the computer may solve this problem.) Be sure, also, to report if you're on some kind of medication, as many types of medication can affect eyesight.

Chapter 6

A Self-Defense Primer

Workspace Defenses

The previous chapters contain information that can enable you to make adjustments in your work area, and to raise issues with your supervisor about those aspects that you can't control. The pictures and some of the diagrams illustrate impractical, and in some cases hurtful, practices and positions, and we've tried to provide guidelines for more functional, and safer, strategies for your workspace setup.

The office workspace is complex, given the tremendous variety in human measurements and the many different types of equipment: chairs, tables and worktops, desks, keyboards, monitors, telephones, documents of different sorts, lights, room configurations, and so forth. If you've taken careful note of all these variables, including your own body measurements and your own history of injuries, you can begin to work on setup and balance issues that are unique to your job. You should write these down, whether they involve lighting, desktop layout, or whatever. Your aim should be, with the aid of this book, to become your own front-line ergonomics analyst. Once you've done your assessment, you can intelligently discuss your workplace situation with your fellow workers, your supervisor, facilities managers, and any ergonomic consultants who may be doing assessments.

You will know what new equipment you may need, and you will be able to defend this need in a professional way.

Likewise, if you are a supervisor or purchaser, you can take a fresh look at the present situation in your offices. You can give thought to problems you detect, set priorities, and devise plans for training and improvements. It is unlikely that either you or your employees can correct everything by yourselves; if you need further assistance from a consultant, you can ask good questions and play an active part in the process of evaluation, and can help shape recommendations to upper management.

The challenge and the frustration of office ergonomics is that what is a good solution for one person may be of no help whatsoever to another. A "one-size-fits-all" approach is simply not possible. This book helps you define what is right for an individual—yourself, or perhaps a fellow employee, if you are a supervisor—and now you need to apply this information so that you won't get hurt.

There is always a limit to the budget for equipment and furnishings; every year, too, new options and refinements become available on the market. Also, aspects of your job may change, requiring a fresh look at your setup.

And as the years pass, your body also changes, often accumulating injuries.

So in taking charge of your situation, you'll need to use a number of strategies and techniques to assure that you get the most out of, and give the most to, the aspects of comfort and function in your job. As things change, you will need to repeat the analytic process.

Professional Defenses

Every week Rab Cross sees and treats employees who suffer from RSI. A significant number of injuries arise from misuse of the keyboard. Many patients with wrist and hand problems lack good keyboard skills; had they had such skills, it is likely that the RSI would not have occurred. Among the common bad habits reported or observed by Rab is the use of only the two index fingers to type; this is a throwback to the amateur "hunt-and-peck" approach of the typewriter area. Modern computer work usually involves a much faster-paced work effort, as well as more hours spent at the keyboard, and so using just these two fingers often leads to hand and finger pain.

Another bad habit is failure to use the little fingers in typing; occasionally, the thumb and ring fingers are also exempted. One patient, a professor working on a book, used only four fingers to do the work, the middle and index fingers of each hand. She became disabled with acute tendinitis and had to have another person take over the typing.

Another very common, and very destructive, habit is anchoring the wrists on the desk, stretching the fingers to reach the keys.

All of these bad habits can be undone by simply taking a typing course given by a professional, in an ergonomically suitable setting. We believe it is likely that most computer users have never had any training in proper typing techniques. The investment of a few hours in a course can instill good habits, and help you to avoid wrist, hand, and finger injury.

Physical Defenses

Once your workspace is well set up, and you know how to use your equipment, you're halfway toward the goal of keeping yourself physically healthy while on the job. Here are the four additional things you've got to do to get yourself all the way there. You've already encountered some of these in earlier chapters.

Take Ergobreaks

This is the single most important defense against RSI. Get up from your desk, go for a walk down the hall, do some filing, get yourself something to drink, or whatever, but leave the monitor and keyboard to themselves for a few moments. How often? John Vaughan, a certified ergonomist who works with some of the largest computer companies, says:

> My advice is to do that every thirty to forty minutes. You almost have to have a clock . . . because if you're

> working at a computer, thirty minutes is like a heartbeat; it doesn't register.
>
> The problem with the whole [work] setup is that it's a very static posture. . . . I was in China a couple of years ago, and saw people every morning doing Tai Chi, some very old people as well as very young. Continuing to be flexible is key, so far as adapting yourself to the computer is concerned.

One of the problems with trying to remember to take ergobreaks is that computer work is, frankly, addictive. Somehow, even the worst of business tasks seems easier when it is done on a computer. The results come so quickly that one is pushed to wring even more out of the machine. It is common, therefore, for workers to keep plugging doggedly away, even past the point of fatigue. It takes discipline, and awareness of how important it is to your health, to force yourself out of the chair, and to take a five-minute break.

Stretch

If you think about the position your body is in while you're working, it is probably the following:

- Lower legs hanging down
- Thighs horizontal
- Hips flexed
- Back vertical
- Shoulders rounded
- Head flexed forward
- Upper arms hanging down
- Lower arms horizontal
- Hands moving from side to side
- Fingers moving up and down

If you simply reverse all these positions, you've got the ideal stretch routine for a computer user.

Stretch at your desk or, better, do that *and* do some stretching in a standing position.

What kinds of stretches? There are six basic ones, and they can be done—all of them—in two minutes.

1. *Overall body stretch.* Get out of your chair, raise your arms above your head, hands close together, and slowly reach for the sky!
2. *Shoulder blade stretch.* Clasp your hands together behind your head and try to pinch your shoulder blades together.
3. *Shoulder shrugs.* Slowly shrug your shoulders five times, raising the shoulders as far as is comfortable.
4. *Shoulder rolls.* Slowly roll your shoulders five times forward, then five times backward.
5. *Head tilts.* Slowly tilt your head to the right, stopping when you feel the stretch. Then tilt your head slowly to the left. Repeat two more times.
6. *Wrist exercises.* Figure 6-1 depicts four different exercises:
   —Illustration a: Rest forearm on edge of desk. Grasp fingers of one hand and gently bend back wrist for five seconds.

Figure 6-1. Wrist exercises.

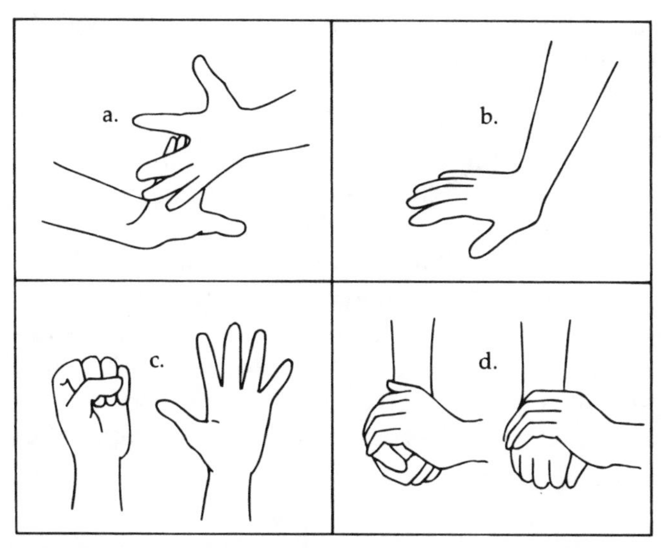

—Illustration b: Gently press hand against desktop, stretching fingers and wrist for five seconds.
—Illustration c: Tightly clench hand and release, fanning out fingers five times.
—Illustration d: For five seconds at a time, make a loose fist and gently press down against clenched hand, keeping the wrist straight with palm up, then palm down.

Do all the stretches listed above. How often? Try to do them five or six times a day, during your ergobreaks or at your desk. Get your computer to remind you! Again, John Vaughan: "When people try some of these stretches, they are just amazed at how much better they feel. It's almost miraculous."

If your company is one of those enlightened employers who allow massage therapists to visit the premises, get a neck, shoulder, and upper back massage every so often. These take about twenty minutes, and produce sheer bliss.

Exercise

Sure, most of us have been meaning to do this for a while now, but we've been just too busy to start. Aerobic exercise three times a week, for a minimum of half an hour, makes a big difference in our strength and stamina on the job. Strengthening our back and upper body muscles enables us to better withstand the stresses of the workplace.

Also, people who exercise tend to get through the first part of the week more easily than the others. Just ask them! So join that company health club, or the fancy workout place down the street. If you feel you don't look so good in snazzy stretch fabrics, wear sweatpants—who cares? If you had an injury years ago and it's kept you off the exercise circuit, consult a physical therapist or specialist in sports medicine to find out what you can do. If you're age 50 or 60 or older and think it's too late to get started, think again. There are groups, coaches, teams that are open only to older people, and that welcome absolute beginners—and that can help you get into superb shape in just a few months.

Adjust Your Chair

This is another tip from John Vaughan. He says:

> Adjust the chair. When you're seated and you have the back straight up, and you're sitting very erect, you'll feel great for a while. You're supporting your upper torso with your lower back muscles. As you tire you're going to start to slump, leaning forward . . . and that puts other strains on your neck; you start to hunch your shoulders . . . to compensate.
>
> If you'll just adjust the back of the chair, moving it backwards, you'll be resting the lower back muscles; they're now being supported by the chair. . . . Change that back several times a day.

Emotional Defenses

Just as some of us tend to take the problems of the office home with us at night, so many of us bring personal problems into the workplace. When the stress from these "outside" problems is added to that of the normal workday, particularly in an ergonomically unsound environment, the consequences can be disastrous, not only to work performance, but also to our physical health.

A brief summary of how stress can affect the body's physical well-being is set forth in an interesting book by Ray C. Wunderlich, M.D. He describes stress as "*dys*-stress," "the state of harassment that results when one is unhappy, unfulfilled, worried, uncertain, overburdened, fearful, guilty, underexercised, underrested, and generally malcontent."*

Wunderlich relates stress to tight muscles, a resultant lack of blood supply, and an "inordinate buildup of waste metabolic products . . . in tissues."† This in turn can be related to the failure

***The Natural Treatment of Carpal Tunnel Syndrome*, New Canaan, Conn.: Keats Publishing, 1993.

†Ibid.

of a local disorder, such as carpal tunnel syndrome, to respond well to therapies.

Here are four actual case histories of people who imported stress from their outside lives into the workplace (the names have been changed) that illustrate the point. In these cases, the stresses may not have directly contributed to or worsened the physical condition, but their effect on the employees was such that the end result was worse physical trauma than would otherwise have occurred.

1. Kathy, age 44, had been working long hours each day for about half a year. Her husband, an engineer, also in his mid-forties, had been laid off the previous summer, and still hadn't found a job. He was sullen, stayed in the house most of the day, had given up his search for work, and watched a lot of television.

Financial pressures were such that Christmas presents for the two children were out of the question, Kathy felt. To bring in money for the holidays, Kathy suggested to her husband that he get a job as a department store Santa; he flew into a rage.

Kathy worked as a customer service phone representative, logging problems and complaints into her computer. She had a new boss, and was trying to make a good impression. She got severe backaches and headaches at work, and reported feeling overwhelmed. In particular, she felt unable to be cheerful on the phone to customers, even though she knew that this was one of the most important aspects of her job. After all, she was the only direct contact in the company for most of these callers.

Kathy's unhappy financial and home situation so preoccupied her that she was not aware that her office chair was actually the culprit causing the headaches and sore back. She had assumed that these were just part of her general feelings of depression. The badly fitting chair was discovered during an "ergonomic audit" of the workplace. Its replacement, along with some stretching, resulted in Kathy's regaining a feeling that she controlled her destiny in the workplace.

2. Charles, age 47, is single. His mother lives with him, and recently was diagnosed as having cancer. Three relatives died of the same form of cancer.

Charles is an accounting supervisor, and spends many evenings preparing reports. He uses his computer, a calculator, and the number pad on his keyboard extensively. Over the past few months, he has developed an ache in his right shoulder that lasts through the night. He finds it difficult, and at times impossible, to lift his mother, though this used to be fairly easy. Some mornings, driving to work, he experiences shoulder pain.

He has the feeling that no one cares about his troubles.

3. Sue, age 37, works in Customer Service and has always been the star performer of her department. Recently, her hands have become numb for varying periods of time, especially at night. She hasn't been able to find the reason for this condition; she is far more preoccupied with the worries she has about her daughter, who lives at home and who has told her parents that she is a lesbian. Mother and daughter have started attending counseling sessions together in an attempt to reestablish communications.

Sue's husband is very angry about the daughter, but will not talk to her about his feelings. He has begun to drink heavily, in bars and at a club.

4. Beth, age 31, has been given her first warning for attendance and performance issues. She has been twelve years in the same department and knows her job well.

She has acute pain in her wrists, which appears to have been induced by a new pattern of behavior: leaning on her forearms against the desk front while pretending to look busy. She takes long bathroom breaks, and her coworkers have found her crying in the bathroom on several occasions.

Beth has told one or two friends that she is consumed with worry about her eleven-year-old daughter, who is doing poorly in school and has started hanging around with an older bunch of kids. Beth's husband is disabled and cannot work.

In the foregoing cases, the employees were preoccupied with personal problems to such an extent that they ignored or overlooked the signs of RSI. In each case, the risk factors were specified and the resulting physical diagnosis was made, thanks to an outside intervention by an ergonomist or by a supervisor.

For each such case where outside intervention provided a solution to the physical problem, there may be many where the emotional issues simply overwhelm the employee, making her or him much more susceptible to injury.

If you have serious problems in your personal life, it is likely that you will either not be aware of workplace dangers or tend to accord them rather low priority. This is an especially harmful state of affairs when you are in financial straits, because it places your ability to earn a salary at considerable risk. It is therefore in your great interest to have safe working conditions, and also to get professional help for your personal problems. The physical and emotional sides of your life are intertwined; both need attention.

Chapter 7

What Organizations Can Do: Three Case Studies

We thought it would be interesting and useful to show what three companies in very different industries are doing to meet the challenges presented by the work environment, in an effort to reduce the suffering and economic drain of RSI. The first of these companies is in the insurance industry. Dr. Rab Cross has served as a consultant to this company, and has been able to gain a unique perspective on how a company reacts, both when it first becomes aware of the impact of RSI on its employees and as its response to this problem is shaped. Dr. Cross presents this case in narrative form. The second company is in the computer industry. Bill Schaffer has worked for it for seven years as a manager, and presents its story through interviews with an official who has the responsibility for devising and implementing a strategy for coping with RSI and its causes.

The third organization actually isn't a company at all. Instead, it's part of the U.S. Army—a research laboratory that does basic and applied research on how to make people (soldiers) more comfortable. The lab has its own staff of "human factors engineers"; we'll learn what happened when they discovered that RSI had made an appearance at the facility.

We hope the examples of these organizations will assist and

encourage managers and workers to carry out their own campaigns to identify, and reduce to the maximum extent possible, the causes of RSI in their own office workplaces.

Phoenix Home Life Mutual Insurance Company

Phoenix Home Life is the nation's thirteenth largest mutual insurance company. It provides a wide range of financial services, including individual life insurance and annuities; group life, health, and disability insurance; mutual funds; real estate management; and reinsurance. The company has 3,600 employees in four home office locations in New England and New York, as well as dozens of field offices nationwide. The focus of the RSI problem was in the Greenfield, Massachusetts, facility, where many operational functions are performed, including paying medical and dental claims, servicing policies, and processing mutual fund transactions.

Beginning in late 1991, employees in the Greenfield office began reporting symptoms of RSI. Dr. Cross, who had provided occupational medical services to Phoenix from 1984 to 1991, could recall no instances of physical problems related to the office environment prior to 1991. In retrospect, it was noted that in the late 1980s, the workflow in Greenfield was "streamlined" so that employees tended to handle more specialized tasks. It was also during this time period that computer terminals appeared on nearly all employees' desks as processes were converted from paper to computer.

In 1992, the company responded to RSI by forming a committee to study the problem; the committee included the occupational health nurses and the corporate risk manager. Employee awareness and education efforts were initiated, including distribution of booklets containing information about RSI and ergonomic approaches aimed at preventing it. A video was created, illustrating proper workstation ergonomics and stretching exercises to be done during regular work breaks. Working in cooperation with the company's workers' compensation carrier, the occupational health nurses and the facilities staff were trained in the techniques of workstation evaluation, and identified sources of adaptive equipment, including wrist rests, keyboard trays, lumbar pillows, and the like.

Despite these interventions, the number of employees with new RSI claims, as well as the number of lost and restricted days, continued to increase dramatically in 1993. Productivity dropped as some departments tried to cope with up to 10 percent of the workforce on restricted duty. In late 1993, Ian Bowers, vice president of human resources, and Rob Kinney, the company's medical director, contacted Dr. Cross for assistance in coping with the RSI problems in Greenfield.

After consultation with senior management, the company contracted with Dr. Cross for a facilitywide evaluation of the workspaces and equipment, making liberal use of photography to document findings. Dr. Cross also interviewed a number of injured employees and conducted physical examinations.

The injuries suffered by the employees were very real and, in some cases, quite severe. Data analysis and evaluation over a period of months revealed that most injuries were clearly work-related or aggravated by work, and were centered around the use of computers. There was no single culprit and no single solution. In particular, it was not the computers that were to blame, but rather the way they were being used, as well as a multitude of environmental factors.

Because of his expertise in both the pathophysiology of RSI and the ergonomic approaches to remedying it, Dr. Cross was able to suggest changes in the work environment to help reduce RSI problems for the Greenfield employees. However, it became evident that it would not be possible to provide an optimal ergonomic environment using the existing office furniture, which consisted of nonadjustable work surfaces and chairs with limited adjustability. In a presentation to management, Dr. Cross showed slides of typical workspace problems, including some of the attempts to improve the workstation with various types of adaptive equipment. Combining an introductory course in office ergonomics with a vivid demonstration of employees in nonergonomic settings, Dr. Cross was able to mount a persuasive argument in favor of upgrading the office furniture.

The price tag for the project, estimated at $3 million, required careful consideration by senior management. However, the RSI problem was clearly hurting the company, as well as the individual employees. Productivity and morale were falling, and workers' compensation costs and the costs of overtime and temporary workers were increasing. In the final analysis, the decision to go ahead with the new office furniture made sense from a business, as well as a humanitarian, standpoint.

Task Force

A task force was designated to study the issue further and produce a detailed plan. The task force was composed of the facilities manager, the purchasing agent, the vice president of administrative services, the occupational health nurse, and Dr. Cross. These people decided that the easiest way to get a rapid handle on possible solutions was to visit NEOCON, a yearly office furniture and equipment trade show in Chicago. So that no time would be wasted, evaluation forms for furniture were developed in advance, to enable systematic gathering of data. In June 1994 the team went to Chicago.

At the show, the team sat in dozens of different chairs and learned to work the various adjustments; it did the same with desks and systems furniture. It was much more "real world" than simply viewing all these items in a catalog, which is often the way office furniture is selected. Phoenix's facilities manager was particularly interested in how workspace structural elements would fit into the space the company occupied.

Of course, all the relevant cost information was also gathered. The team met each evening to discuss the pros and cons of various models.

During the summer, the team continued to meet regularly and continued its evaluation of the many models of chairs and other furniture.

Chair Replacement

Finally, three chair manufacturers were selected as suppliers. These three manufacturers collectively offered a wide range of chairs to accommodate the great range of employees' heights, weights, widths, and shapes—a range that was evident from the photographic record, part of the initial ergonomic survey.

Phoenix eventually purchased two chairs of each of the models from the manufacturers. Early in 1995 a volunteer team from different company departments was trained in the controls of all of the models. These volunteers were to work with employees to explain the adjustment features, and to help each employee find the best chair match.

The following week, in groups of six, each employee in the company received a personal fitting session, usually lasting fifteen minutes, but sometimes going as long as an hour. The facilities manager, the occupational health nurse, the fitting team, and Dr. Cross were in attendance for the entire "fit-sit" program, which lasted three days. Every employee was able to find a chair that suited his or her particular needs.

Work Surface Replacement

During the summer, the task force had also selected two manufacturers of work surfaces. The principal criterion for selection was that the surfaces had to be adjustable by the employee. The two manufacturers provided units for evaluation, one 6′ by 6′, and the other 6′ by 8′. The corner of each L-shaped unit was used for placement of the monitor, enabling proper alignment of the monitor screen and keyboard and proper distance from the employee.

This time, departmental managers, supervisors, and many employees inspected the units to see if the work they had to do could, in fact, be done with that furniture. Of course, the crank adjustment mechanisms were carefully evaluated by the facilities people, as were wiring, panel materials, types of cabinets, and overall construction. Ultimately the selection was made following a competitive bidding process. The chairs were purchased first, one for each employee, along with several extra units of each model, in case employees changed their minds, and also to accommodate changes in the workforce. The work surface units were phased in over a nine-month period.

Other Actions

Many issues remained to be resolved. Phoenix instituted a Keyboard Proficiency Training Program, having discovered that many of its employees were not keyboard-trained touch typists. Many had developed injurious habits. Some of these, in the vernacular of the ergonomist, are space-bar thumb, pop-up pinkie, finger splays, and key pounding. These are now being addressed at Phoenix in an ongoing fashion.

Another relearning that had to occur involved proper use of the chair. Employees often didn't bother to adjust or readjust their chairs. They were not used to the new features and had to develop the habit of using them. And their new chairs go with them if they switch location or job.

Employees at Phoenix have another very important responsibility. This is to stretch and strengthen their upper bodies, so as to counteract the strain that office work—even correctly carried out—places on the arms, neck, shoulders, and other areas we talk about in Chapter 4. Studies have shown that many employees, and females in particular, have minimal upper body strength. Medical examinations at Phoenix have detected a high degree of muscle shortening and low-grade tendinitis in the undermuscled shoulder and lower back.

Demonstrations and training for body stretching, motion, flexibility, and strengthening are being provided at Phoenix. In the last analysis, however, it is the individual employee commitment that will determine a successful outcome.

The company's vice president for human resources, Ian Bowers, made the following observations about RSI:

> It is clear from our efforts over the past several years that a company ignores its environmental issues and training employees in the proper use of equipment at its peril. The success we have had in reducing the incidence and severity of RSI conditions speaks volumes for proactive responses to trends. If we had ignored the data, our employees and the company's bottom line would have continued to suffer.

Sun Microsystems, Inc.

Sun Microsystems is a computer and software company located in Mountain View, California. Founded in 1982, the company grew rapidly, and is now a worldwide force in the high-tech industry. The company has over 13,000 employees worldwide. Each employee has a powerful computer on his or her desk or work surface, and each computer is linked to all the others in a worldwide network.

The following is adapted from an interview with Dennis Mattison, who is Sun's corporate environmental health and safety manager.

Q: *How did Sun get started on its ergonomic program?*

A: Sun has always had a culture where it tries to do the best it can to enhance employee quality of life. The idea is, when employees are happy, employees are more productive. We actually designed the program around making sure that employees had comfortable seating, comfortable positions for their workstations. Many employees at Sun sit in front of workstations all day long. So it was mostly driven by comfort.

There were a few issues arising about injuries, so I went to our workers' compensation department and talked about collecting data to take a look at them. It turned out that a lot of employees in the high-tech and other industries sit in front of workstations all day, and there were increasing claims of some upper-extremity cumulative trauma. There weren't any regulations at the time, and in fact there still aren't any, whether at the federal level or from the state of California. We decided to go ahead and begin forming what we call our employee ergonomic program, to address some of the needs for proper seating and that sort of thing.

Q: *Is it fair to say that a comfortable employee is in rather good shape, ergonomically speaking?*

A: If employees feel they're comfortable, they are probably a long way toward being ergonomically fit.

Q: *Are there some data that show a correlation between the comfort of an employee and ergonomically correct surroundings, and productivity?*

A: Productivity is a hard thing to measure. The way we measure it is through lost-time days.

Q: *How about workers' compensation claims?*

A: You can file a claim and not be out of work at all. You can just go see a doctor and be put on some program at work. But you wouldn't have to leave work. So I think the key thing, for line managers, at least, is lost time. That hits them directly.

Q: *What is the content of Sun's program at this time?*

A: It starts with education. We've got a video, eight minutes long—in the Sun style, it's short, fast, and to the point. It's designed for use at a staff meeting, for example. The purpose is to pique interest. It shows a Sun employee being properly set up, and fitted with a good ergonomic position at a workstation.

We also have a pamphlet for employees in the United States and overseas; it's in six different languages. And we have a pamphlet that goes out with each computer we deliver to a customer.

Then, on line, we have a large Environmental Health and Safety volume, of which about forty-five pages are devoted to upper-extremity trauma, or RSI. It discusses the history of RSI, talks about the Sun program, and describes some of the equipment employees can get.

Q: *Do you have training for employees?*

A: Yes. A training session is about forty-five minutes long. We take a person from the audience and set him or her up with a mock Sun workstation. We show how to sit properly. The idea is to get the people in the audience to go back to their workplaces and make some intelligent decisions on how they should sit and the organization of the workspace. The whole program is designed around employee involvement and awareness. We feel that most employees at Sun don't have a significant issue with RSI.

Q: *I notice that among the materials here, there is a catalog that shows a lot of equipment. So the ergonomics program encourages employees to order what they need?*

A: Yes. We also have a demo center in each of our major campuses where employees can go, and sit down and try all the various chairs and accessories—all the equipment they can order from that catalog. Twice a year we take a look at new ergonomic equipment and supplies. Ergonomic standards change a lot; people invent new things. The employees in our beta test have a big say on whether this stuff actually gets into our catalog.

Q: *How many training sessions have you given yourself?*

A: About 200 in the last two years, throughout the United States and Europe, primarily. We were sometimes doing four or five a week.

Q: *Is top management aware of the potential problems and what the company is doing?*

A: Sun is a self-empowered company, a midlevel management company. We had to get top management buy-in for the initial money invested in the program. But as for whether or not they're involved in all the facets of the program, I don't think so, unless they've used it themselves.

Q: *Would you say that ergonomics is a continuing area of concern?*

A: If employees sit right and work right, I'm not sure it's a problem. If they learn up front, and get the proper literature, and begin properly, I'm not sure it's a problem. I feel strongly that it's not the equipment that is the issue. It's how employees use the equipment.

I know people who work at their workstations all day long and never have an issue about this. We didn't start our program as a response to problems; we started it as a quality-of-life issue. Sun is a very proactive company, and that was our driving force. The program was well under way before we started getting data on the [RSI-related] results.

Q: *How did you measure it?*

A: We went through our workers' compensation department and got figures from them on cost per claim for upper-extremity cumulative trauma. We found that the cost per claim has been going down yearly. The number of claims has been going up, because more employees are hearing about this. We expect that the number of claims will also taper off. We'll eventually have a low cost per claim and a low number of claims, down to some level. Sun has an aging workforce, and so we don't expect the claims ever to go totally away.

Q: *You've mentioned correct seating as being especially important.*

A: It's an important part of the total surroundings. If you take a look at our new Menlo Park campus, you'll see that each employee has adjustable work surfaces. So there can be one height for the keyboard and one for writing. We also have a special chair there that adjusts in almost every way you can think of. Plus, the lighting is all indirect. It's a really good step for Sun, and we pushed very hard to get the adjustable furniture.

Q: *Do you have any thoughts you want to share with the managers or employees of smaller companies that don't have an ergonomics program?*

A: First off, companies have to realize and understand the importance of quality of life and productivity. They also have to understand that an injured employee isn't much of an asset to the company! If they delved into their compensation figures, they would probably find that an ergonomics program would save them money, not cost them money.

I think a lot of companies have the idea that this is going to be costly, but don't actually do any calculations. If they did, I think they'd be surprised at how much a program would save them. You don't have to go all out, the way Sun does. Just getting the information across to the employees is very important.

Many companies have pretty good chairs, but they have to be adjusted, and maybe modified with equipment such as footrests or adjustable keyboard trays. And they need to use task lighting rather than overhead lighting, which might cause eyestrain. You don't have to spend a fortune. You just have to have a direction. And you probably have to have someone who's interested in championing that program and can show it to midlevel management and get them to focus on it. At Sun, we've also got an online bulletin board called ergo@sun, where you can ask a question or request a presentation. It's United States–wide, and it works well for employees. We

have lots of ways employees can call up, reach somebody, talk with someone, ask a question.

U.S. Army RD&E Center, Natick, Massachusetts*

As part of the research for this book, we visited the Human Factors Branch of the Science and Technology Directorate at this research center in New England. There we spoke with Dr. Carolyn Bensel, senior research psychologist; Steven Paquette, physical anthropologist; and Marcie Kronberg, human factors engineer. The Natick center, or Natick, as we refer to it from here on, does basic and applied research to develop clothing, personal equipment, and other equipment for the U.S. Army, so that it will be compatible with the user. We discovered that the Natick facility itself had encountered an RSI problem among its 900 employees. But because of the nature of the research carried out there—research intended to improve the lot of the ordinary soldier—Natick had special resources available to tackle the appearance of RSI.

Dr. Bensel explained for us how "human factor engineering" got its start.

> In the U.S., human factors got its start when experimental psychologists . . . got dragged into the U.S. war effort, and were asked to do things like develop selection tests, or develop tests to try and predict who is going to be worthy of being trained as an aviator. . . .
>
> Immediately prior to our entry into the Second World War there was an interest not only in selection, but also looking more closely at equipment design, say the design of an aircraft cockpit. . . . My predecessors . . . found themselves saying "I'm not an experimental

*The views presented in this section are those of the persons interviewed, and do not necessarily represent the view of the Department of Defense or its components, including the Soldier System Command and the Natick Research, Development and Engineering Center.

> psychologist any more, because I apply too much! . . . So I must be something different." So, human factors came to be.

Marcie Kronberg described how RSI made its appearance at Natick:

> A few years ago there was a woman who had a complaint of a repetitive nature in our procurement area, where they do a lot of the billing. She was using a computer, an adding machine, and talking on the phone a lot.
>
> She had been to the doctor, and actually had worn a neck brace and some things like that. Her boss got in touch with our safety office, who then got in touch with the human factors area, and said he thought we should be looking at this.

This was long before the press and public became aware of the dangers of repetitive stress injuries in the office workplace, and when Ms. Kronberg visited the employee's office, she didn't know that that visit was merely the first step leading toward a preventive program at Natick.

> I went over, and took an anthropometrist with me, in order to get some measurements of the seat, and such. I found whatever information I could, because workspace ergonomics isn't an area I generally dealt with.
>
> I did things like adjusted the seat, and the desk area, and the monitors. Once I got to that one woman and starting looking at her workspace, everybody else was coming over.
>
> [RSI] wasn't being reported through our normal channels, because people didn't see it as a work-related type of thing.

In 1993 a significant number of complaints about work-related injuries began to rise, perhaps because of the attention then just beginning to be paid to the problem by the media. By that time an ergonomics working group had been formed at Natick, and it was already going out to look at people's workspaces. This itself contributed to employees' understanding of how the improper use of equipment could affect their health.

> We targeted a couple of areas we thought would be high risk because of the kinds of work they did. We looked at their office areas and modified them. Some we went to great lengths to modify, especially as to where the computer terminals were sitting . . . we basically had some of the people in the shop building come and reconfigure the consoles, and move everything around. They've ordered a lot of new chairs, some new furniture. . . .
>
> What we pay most attention to is the adjustability of the chairs. [I became knowledgeable] from catalogs, from reading the scientific literature. Some of it came just from seeing the different chairs, testing them out.

RSI-related training at Natick is contracted out; the typical training session takes two hours and involves not only presentations, slides, and handouts, but also practical demonstrations.

> Training is done by bringing in different pieces of furniture, and showing people the correct way to adjust them. [The trainers] do explain the medical aspects of the different diseases, why these things occur, through different kinds of sitting arrangements, and why the muscle strains might occur and lead to problems.

At Natick, as in the other organizations we visited, training led employees to return to their offices and make the changes needed to fix their work environment. If this could be done without purchases, so much the better. As Ms. Kronberg told us, when the monitor must be raised, it's not always necessary to buy a device to do this. But no compromise is made so far as the chair is concerned. Headsets, too, are purchased when employees spend a lot of time on the telephone.

Ms. Kronberg also talked about the importance of taking breaks from the repetitive tasks of everyday work.

> I think it's management's responsibility to accommodate the employees by allowing them to take rest breaks and do the exercises. They need to work that into the work schedules.

And when asked for any final words of wisdom for all the office workers out there, she said,

> Really use the knowledge you're being given. Don't wait until you're in pain, because by then it's too late, and you've got to take some kind of corrective action. Really know where the risk factors are, and if you fit into any of those categories where they are more prevalent, take action. . . . Don't be afraid to ask [for help], especially if you think you may be doing something that can hurt you.

Chapter 8

Back to the Future

The use of computers, and hence the risk of RSI, is spreading far beyond the traditional workplace. For example, a recent study revealed that 37 percent of American households have one or more computers; it was also estimated that their owners spent about thirteen hours each week in front of these systems. Of that time, about 80 percent was related to work activity.

More and more workers are moving out of the office and into "telecommuting"—working at home on a computer that is hooked up to the company's network through a modem. And more and more small businesses that are heavily dependent on computers are being established in the home.

We have looked at many home office setups, and have observed that while people sometimes make a substantial investment in the computer and its peripherals, they often skimp when it comes to the work surface and the chair. It is not uncommon to find a folding chair or a chair snitched from a breakfast area in front of the computer, often with cushions affixed to it with tape or twine.

Lighting, too, is often poorly placed with respect to the monitor.

When companies permit workers to work at home several days a week, or perhaps even full time, they may furnish the computer or workstation, but leave the responsibility for the furniture to the employee. Employees may be reluctant to invest several hundred dollars in an ergonomically correct chair, especially where this may require the approval of a spouse or partner who also has an interest in the household budget.

It is imperative that the best ergonomic environment possible be established in the home office. It is better to save money by buying a less powerful computer, and invest the savings in a good chair and appropriate work surface than to try to cut corners on the furniture.

The home is not the only place besides a business where there lurks a danger of RSI. A recent survey of an Ivy League college revealed that *every* student possessed and used a computer. Some schools even require this.

At the college there was a great variety of types and models; the common thread, however, was that there was an almost total lack of, and disregard for, the most basic ergonomic elements of safety in the computing environment. Chairs were generally too low, and students were usually pressing their wrists and forearms directly onto the sharp edges of desks while using the keyboard and mouse. There will certainly be a wave of injuries caused by these practices in the next few years, unless schools begin immediately to conduct educational campaigns, and students heed them. The single biggest risk is of contracting carpal tunnel syndrome through swelling of wrist tissues caused by pressure on them.

Another risk is to eyesight, as students (and adults in the home) become addicted to the new wave of computer games. Game players spend hours staring at the screen and, of course, at the keyboard and mouse or joystick. And use of the Internet is spreading rapidly, as millions of new users "get wired."

Laptop computer use is growing rapidly among businesspeople, designers, lawyers, writers, and students. Laptops are a common sight in cafes, on airplanes, in libraries, and even in university classrooms and laboratories. Laptops pose no significant risk of RSI when used for brief periods. Problems arise, however, when they are used for report writing. The screen sits directly behind the keyboard (usually—some models permit a detached screen), and the keyboard is smaller than the full-sized desktop models.* The screen placement often induces a contorted posture; either the neck and upper back must bend far forward to permit viewing the screen, or else the arms are held high and the wrists press against the edge of the keyboard, while the neck and shoulders are close to neutral.

*IBM's "Butterfly" model, with a two-part, swing-out, full-size keyboard, is without doubt pointing the way all laptops will go in the next few years.

The best way to use a laptop while traveling, or otherwise out of the office, is intermittently. At home or at work, use a docking station, or attach a full-size keyboard and use a separate monitor; most good laptops allow these to be attached.

Chapter 9

If You Think You Have RSI

Failure to recognize a medical problem in its early stages is one of the main reasons people end up requiring therapy, medication, time off, and even surgery. Therefore, it is important that you be aware of how you feel, and that you report problems promptly. The early signs, as you now know, are undue fatigue, pains, or tingling and numbness.

The first step is to tell your supervisor. He or she has the responsibility to gather the story from you, so that an incident report can be completed. You needn't go into medical details—keep them for your doctor. But let your supervisor or manager know some history of what hurts, when it hurts, how long it's been hurting, and whether you connect the problem to your job in any way.

This reporting procedure is absolutely required for the company's medical insurance (worker's compensation) carrier to become involved, so that bills can be paid. Reporting is also important to let the company know what is going wrong in the workplace, so that it can take remedial action. There may be a general problem; so your reporting can help others, as well. Reporting is also usually an OSHA requirement for a company.

The next step is to get a good medical evaluation. There are several possible ways to accomplish this:

- Your company may have a professional—a nurse or physician—on site whose role is to interview and examine employees with complaints or problems, to determine the nature of the problem, and to make recommendations

about treatment. This person should be reasonably familiar with your job duties and risks.
- Your company may contract with a doctor off premises to perform the same services for employees. Here again, it is likely that the doctor will have some familiarity with your job tasks.
- You may prefer to see your personal physician. This doctor may or may not be familiar with your work and office setting.

With any of these alternatives, you should expect to have an appropriate examination. Many of the complaints arising from the office setting involve the upper extremities: hands, wrists, elbows, or shoulders. Your symptoms can sometimes mislead both you and the doctor because shooting or fleeting pains or tingling can involve a lot of arm territory. This may make it impossible for you to describe the real locus of the problem accurately. Thus it has become standard medical practice to examine the entire upper body, including the neck, since everything is neurologically connected. If you have a problem on your right side, don't be surprised if the doctor spends a lot of time examining the other side as well, for comparison's sake.

If you don't get this kind of thorough examination, you should find another doctor, one who is familiar with RSI.

During the medical examination, it is not unlikely that places will hurt that you didn't know were even a problem; this is very common with RSI. And some of the examination may be uncomfortable, since the doctor will have to do some probing and stretching of tissues and limbs. It is possible that X rays or blood tests may be recommended.

There may be a time limit on your visit to the doctor—for example, if you use an HMO. It is not unusual for a doctor to send you on to a skilled physical therapist or occupational (hand) therapist for a more complete examination. Many qualified therapists do an excellent job in this capacity. If you feel pretty well poked over by the end of the examination, chances are that a good job has been done.

Treatment for RSI is never as quick as people expect. "Simple" hand and arm problems can drag on for weeks and months after being diagnosed. Remember that as RSI develops, much of the injury occurs silently, below your pain threshold—so there's almost always more there than you expect to find, given the level of pain you're experiencing. Despite the best methods used by therapists, and despite the new drugs that are being used, the healing process requires time—sometimes lots of it. RSI tests the patience of the patient, the supervisor or manager, and often the insurance carrier.

When they are reported early, these sorts of injuries usually get better, and usually don't come back. If you do your part, then the doctors and therapists will do their part. Your body will do its part, and the episode will become history. Of course, you'll have identified the causes of your originally getting into trouble, discussed them with your supervisor, and made the necessary changes to your work area.

If, on the other hand, you develop very serious problems, you will require a specialist with skill and experience in treating these sorts of injuries. This is where you must become a careful shopper for medical care. Here are some suggestions that can make things easier for you:

- Inquire of your therapist where he or she would go. If you promise confidentiality, you are likely to get some good referrals on the spot.
- Ask your company nurse or human resources person about the specialists used in similar cases. Ask fellow employees about their experiences with specialists.
- Ask the insurance carrier for a referral. The insurance company has a strong interest in your early and complete recovery; it can probably give you a list of preferred providers for RSI treatment.
- You can call your local hospital or city or county medical society to obtain the names of orthopedists and hand surgeons in your area. You probably won't get a specific recommendation, but you will get a list of names to work with.
- If you still feel lost or inadequately served, we've included a list of names of organizations dealing with RSI,

along with phone numbers. These names appear in numerous articles and publications about RSI.

Prevention of RSI is truly the best medicine. *ErgoWise* is your key to prevention in the office environment. We hope it will unlock the doors to a successful and healthy future.

Appendix

Following is a list of organizations that provide information, support, and advice concerning repetitive strain injuries, including carpal tunnel syndrome. New organizations and committees are forming all the time and may not be included in this appendix. Your physician will also be able to assist you.

Committees on Occupational Safety and Health (COSH)

COSH groups are nonprofit coalitions consisting of unions and individuals concerned with protecting the safety and health of workers. Your local COSH group distributes books and manuals about repetitive motion disorders.

ALASKA
Alaska Health Project
1818 W. Northern Light Boulevard
Anchorage, AK 99517
907-276-2864/Fax: 907-279-3089

CALIFORNIA
Worksafe/Francis Schreiberg
c/o San Francisco Labor Council
660 Howard Street, 3rd Floor
San Francisco, CA 94105
415-543-2699/Fax: 415-433-5077

LACOSH (Los Angeles COSH)
5855 Venice Boulevard
Los Angeles, CA 90019
213-931-9000/Fax: 213-931-2255

SACOSH (Sacramento COSH)
c/o Fire Fighters, Local 522
3101 Stockton Boulevard
Sacramento, CA 95820
916-442-4390/Fax: 916-446-3057

SCCOSH (Santa Clara COSH)
760 North 1st Street
San Jose, CA 95112
408-998-4050/Fax: 408-998-4051

CONNECTICUT
ConnectiCOSH (Connecticut COSH)
77 Huyshoup Avenue, 2nd floor
Hartford, CT 06106
203-549-1877/Fax: 203-251-6049

DISTRICT OF COLUMBIA
Alice Hamilton Occupational Health Center
410 Seventh Street, SE
Washington, DC 20003

202-543-0005 (DC)/301-731-8530 (MD)
Fax: 202-546-2331/301-731-4142

ILLINOIS
CACOSH (Chicago Area COSH)
37 South Ashland
Chicago, IL 60607
312-666-1611/Fax: 312-243-0492

MAINE
Maine Labor Group on Health
Box V
Augusta, ME 04330
207-622-7823/Fax: 207-622-3483

MASSACHUSETTS
MassCOSH (Massachusetts COSH)
555 Amory Street
Boston, MA 02130
617-524-6686/Fax: 617-524-3508

Western MassCOSH
458 Bridge Street
Springfield, MA 01103
413-731-0760/Fax: 413-732-1881

MICHIGAN
SEMCOSH (Southeast Michigan COSH)
2727 Second Street
Detroit, MI 48206
313-961-3345/Fax: 313-961-3588

MINNESOTA
MN-COSH
c/o Lyle Krych M330
FMC Corp. Naval System Division
4800 East River Road
Minneapolis, MN 55421
612-572-6997/Fax: 612-572-9826

NEW HAMPSHIRE
NHCOSH
110 Sheep Davis Road
Pembroke, NH 03275
603-226-0516/Fax: 603-225-1956

NEW YORK
ALCOSH (Alleghany COSH)
100 East Second Street
Jamestown, NY 14701
716-488-0720

CNYCOSH (Central New York COSH)
615 West Genessee Street
Syracuse, NY 13204
315-471-6187/Fax: 315-422-6514

ENYCOSH (Eastern New York COSH)
c/o Larry Rafferty
121 Erie Boulevard
Schenectady, NY 12305
518-372-4308/Fax: 518-393-3040

NYCOSH (New York COSH)
275 Seventh Avenue, 8th Floor
New York, NY 10001
212-627-3900/Fax: 212-627-9812
914-939-5612 (Lower Hudson)
516-273-1234 (Long Island)

ROCOSH (Rochester COSH)
46 Prince Street
Rochester, NY 14607
716-244-0420

WYNCOSH (Western New York COSH)
2495 Main Street, Suite 438
Buffalo, NY 14214
716-833-5416/Fax: 716-833-7507

NORTH CAROLINA
NCOSH (North Carolina COSH)
P.O. Box 2514
Durham, NC 27715
919-286-9249/Fax: 919-286-4857

OREGON
c/o Dick Edgington
ICWU—Portland

7440 SW 87 Street
Portland, OR 07223
503-244-8429

PENNSYLVANIA
PhilaCOSH (Philadelphia COSH)
3001 Walnut Street, 5th Floor
Philadelphia, PA 19104
215-386-7000/Fax: 215-386-3529

RHODE ISLAND
RICOSH (Rhode Island COSH)
741 Westminster Street
Providence, RI 02903
401-751-2015

TENNESSEE
TNCOSH (Tennessee COSH)
309 Whitecrest Drive
Maryville, TN 37801
615-983-7864

TEXAS
TexCOSH
c/o Karyl Dunson
5735 Regina
Beaumont, TX 77706
409-898-1427

WASHINGTON
WASHCOSH
6770 East Marginal Way South
Seattle, WA 98108
206-767-7426/Fax: 206-762-6433

WISCONSIN
WisCOSH (Wisconsin COSH)
734 North 26th Street
Milwaukee, WI 53230
414-933-2338

ONTARIO, CANADA
WCOSH (Windsor COSH)
547 Victoria Avenue
Windsor, Ontario N9A 4N1
519-254-5157/Fax: 519-254-4192

COSH-Related Groups

CALIFORNIA
Labor Occupational Health Program
2515 Channing Way
Berkeley, CA 94720
510-642-5507/Fax: 510-643-5698

DISTRICT OF COLUMBIA
Occupational Health Foundation
1126 16th Street NW, Room 403
Washington, DC 20036
202-887-1980/Fax: 202-887-0191

NEW JERSEY
New Jersey Work Environment Council
452 East Third Street
Moorestown, NJ 08057
609-886-9405/Fax: 609-866-9708

NEW YORK
Tompkins Cortland Labor Coalition
109 West State Street
Ithaca, NY 14850
607-277-5670

OHIO
Greater Cincinnati Occupational Health Center
10475 Reading Road
Cincinnati, OH 45241
513-769-0561

WEST VIRGINIA
Institute of Labor Studies
710 Knapp Hall
West Virginia University
Morgantown, WV 26506
304-293-3323/Fax: 304-293-7163

Other Sources of Information About RSI

9to5 National Association of Working Women

9to5 provides literature on computer health and safety. 9to5 is a membership organization that works to improve working conditions for women.

9to5
614 Superior Avenue, NW
Cleveland, OH 44113
Hotline: 1-800-522-0925

Office Technology Education Project

This organization offers safety seminars and preventive on-site workshops. It also provides fact sheets about RSI.

1 Summer Street
Somerville, MA 02143
617-776-2777

Association for Repetitive Motion Syndrome

The Association for Repetitive Motion Syndrome publishes a newsletter and distributes information about RSI.

P.O. Box 514
Santa Rosa, CA 95402
Contact: Stephanie Barnes
707-571-0397

RSI Support Groups

Following is a list of RSI support groups. Because of the growing numbers of people being affected by repetitive strain injuries, new support groups are being formed all the time and may not be included.

CALIFORNIA

East Bay RSI Support Group
Contact: Joan Lichterman
510-653-1802

RSI Support Group of San Francisco
Contact: Judy Doane
415-474-7060

Santa Rosa RSI Group
Contact: Stephanie Barnes
707-571-0397

Marin RSI Support Group
Contact: Liza Smith
415-459-0510

SF Peninsula RSI Support Group
Caremark Peninsula Athlete's Center
216 Mosswood Way
South San Francisco, CA 94080
Contact: Lynda Jensen
415-589-0600

CONNECTICUT

The Connecticut Chronic Pain Outreach Network, Inc.
P.O. Box 388
Hartford, CT 06141-0388

NEW YORK

RSI Support Group
Mount Sinai—Irving J. Selikoff Occupational Health Clinical Center
P.O. Box 1252
One Gustave L. Levy Place
New York, NY 10029-6574
Contact: Susan Nobel, M.S.W.
212-241-1527

Government Agency

National Institute of Occupational Safety and Health (NIOSH)

Technical Information Center
4676 Columbia Parkway
Cincinnati, OH 45226
1-800-356-4674/Fax: 513-533-8573

On-Line Information

Tapping into the World Wide Web allows you to punch up repetitive strain injury and read all about preventive measures and other up-to-the-minute information concerning RSI. America Online has a section in the reference section under the key "health" called "Better Health and Medical Forum" that contains RSI self-help group information, as well as a forum offering RSI on-line chats and an on-line meeting schedule.

Suggested Readings

Donkin, Scott W. *Sitting on the Job.* Boston: Houghton Mifflin, 1987.

Mills, Wendy Chalmers. *RSI.* New York: HarperCollins, 1994.

Pascarelli, Emil, and Deborah Quilter. *Repetitive Strain Injury: A Computer User's Guide.* New York: Wiley, 1994.

Putz-Anderson, Vern. *Cumulative Trauma Disorders: A Manual for Musculoskeletal Diseases of the Upper Limbs.* London: Taylor and Francis, 1988.

Sellers, Don. *ZAP! How Your Computer Can Hurt You, and What You Can Do About It.* Berkeley, Calif.: Peachpit Press, 1994.

Index